T0205891

Statistical Models in Toxicology

Statistical Models in Toxicology

Mehdi Razzaghi

CRC Press
Taylor & Francis Group
Boca Raton London New York

CRC Press is an imprint of the
Taylor & Francis Group, an **informa** business

A CHAPMAN & HALL BOOK

"Just as mice and rats are not humans, a mathematical model is not a human, but like mice and rats, a model can be thought to be reprehensive of human responses to chemical exposure."

John Paul Gosling, University of Leeds

Contents

Preface

This book describes the statistical models for the analysis of data from toxicological experiments with a strong emphasis on application. Although the use of statistical methods to analyze toxicological data dates back to the 1930s, the development of models to describe toxicological experiments began in the 1960s and 1970s and continued through the twenty-first century. This book presents some of the most important and commonly used models for the characterization and analysis of data from toxicological experiments. Given the magnitude of literature, there was no hope in covering all models that have been discussed. Rather, the aim was to present statistical models which can provide a basis for understanding the underlying framework of different types of experiments. Thus, the book is not meant to be an exhaustive treatment of all statistical applications and models in toxicology. There are a few excellent books that focus on the applications of statistical methods in toxicology. To name a few, the seminal work of David Salzburg (*Statistics for Toxicologists*) in the 1980s and the more recent book by Ludwig Hothorn (*Statistics in Toxicology Using R*) can be mentioned. In addition, several books with collections of chapters from various authors such as Byron Morgan's *Statistics in Toxicology* have been published in the past. However, to my knowledge a book specifically concentrating on statistical models with real data examples as well as intuitive explanation and derivation of the models and their interpretations has previously not been presented.

The initial motivation for writing this book stemmed from a course by the same title that I taught at the University of Warsaw in Poland while I was on leave during the academic year 2014–2015. The success of the course and enthusiastic reactions from my students encouraged me to assemble the notes that I developed for that course and expand them into a textbook. The technical level of the book is such that it could be used as a textbook in a course for advanced undergraduate and beginning graduate students of statistics, mathematics, and closely related fields. Therefore, it is assumed that the reader is familiar with calculus, matrix algebra, statistical distributions, and statistical inference. Knowledge of linear models would be beneficial, but not necessary. Thus, the book may also be used by students and researchers in toxicology, pharmacology, and public health with the right background. Many parts of the book, however, are written at a more elementary level, amenable to students with the basic knowledge of mathematics and statistics. The goal is to present the broad spectrum of modeling problems in toxicology.

The book is organized into seven chapters. Apart from Chapters 1 and 2 that provide some background, each of the remaining chapters deals with models for a specific type of toxicological experiment and the analysis of outcomes from those experiments. This gives the reader the capability to maneuver through different chapters, and in the classroom setting, it gives the instructor the ability to select chapters as deemed necessary. We begin in Chapter 1 by presenting some basic concepts and definitions in toxicology, branches of toxicology, and some historical notes on the growth of statistical applications in toxicology. Chapter 2 provides the foundational structure for risk assessment and specifically discusses the use of animal bioassay data in health risk assessment. Chapter 3 is devoted to models that characterize the assessment of the effect of a mixture of chemical compounds. Chapters 4 through 7, present, respectively, the statistical models in animal carcinogenicity experiments, developmental toxicity studies, neurotoxicity studies, and finally developmental neurotoxicity experiments. An attempt has been made to provide examples using real data sets from toxicological experiments to demonstrate the application of the presented models. Since many research papers provide summary statistics rather than the actual data set, when the actual experimental data set was needed and not available, the summary statistics were used to generate a typical representative data set for illustrative purposes. The procedure and the generated data sets are presented in Appendixes 1 and 2. The model final exam questions given in Appendix 3 provide a guideline for the instructor and a good practice for the student. The statistical software R was used for calculations and model fitting in many of the examples throughout the book. R is currently the most popular software which is used by a vast majority of researchers and students. It is considered to be a leader and is increasingly utilized by the government and industry. In addition, R can be downloaded for free from the internet. Although R programs and codes are not included in the text, the specific functions in R used to solve each example are stated. Interested readers may contact the author to receive the codes. At the end of each chapter, a set of exercises is given. These exercises are generally a balanced mixture of theoretical and applied problems and provide a deeper understanding of the applications of models presented in that chapter.

The key features of this book can therefore be summarized as follows:

- It is a cross-disciplinary book and can appeal to audiences from different fields and different backgrounds.
- Most of the chapters are independent and can be taught out of order.
- The models presented in the book are based on well-documented research papers. References are provided in the text, and an extensive bibliography is presented at the end of each chapter.
- Every effort has been made to utilize uniform notation throughout the book.
- Depending on the background of the audience, material can generally be covered in a one-semester course.

- Several relevant examples with real experimental data are used to illustrate the application of various models.
- Exercises provide a good mix of theory and application. The goal of these exercises is to help instructors in assigning homework problems and, in turn, give the reader the opportunity to practice the techniques that are discussed in the book.
- The application of the models discussed in different chapters is demonstrated using the statistical software R, which is free and is widely used in statistical research.

My interest in the field of toxicology was initiated through a summer research fellowship from the Oak Ridge Institute for Science and Education (ORISE) at the National Center for Toxicological Research (NCTR) facility of the Food and Drug Administration (FDA) back in the early 1990s. I benefited immensely from my several years of collaboration with the statisticians and scientists at the center that continued through 2005. I am grateful to all of them, specially David Gaylor, Ralph Kodell, Robert Delongchamp, and James Chen. I would also like to acknowledge the fact that my sabbatical leave at the University of Warsaw in Poland that gave me the opportunity to teach a course and develop notes that became a nucleus for this book was supported by a Fulbright grant for which I am grateful. I would also like to thank my colleagues and students in the Department of Statistics at the University of Warsaw and my colleagues at Bloomsburg University. Last, but by no means the least, I would like to thank my family. My wife Mehran has been extremely supportive of me throughout my entire career and has made countless sacrifices to help me.

Finally, I would like to mention that this is a first edition of the book. Although I have tried to be very careful in my writing and presentation, and illustration of statistical models that are utilized in toxicology, I am certain that there may be many shortfalls and much can be improved. I would very much welcome and appreciate any feedback that I could receive from colleagues and students.

Author Bio

Dr. Mehdi Razzaghi is Professor of Statistics at Bloomsburg University and has served in this capacity for 31 years. He holds a BS in mathematics from the University of Sussex in England, a BS in Computer Science from Bloomsburg University and a PhD in statistics from the University of London, England which he received in 1977. He held faculty positions at Marburg University in Germany, University of Kentucky, California State University Chico, and Allameh University in Iran before joining Bloomsburg University faculty in 1987. Dr. Razzaghi is the recipient of the Faculty Recognition Award (2003) and the Outstanding Scholarship Award (2012) from Bloomsburg University. During the 2014–2015 academic year, he was a Fulbright fellow at the University Warsaw in Poland. In 2018, he served as a Fulbright Specialist at the Medical University of Silesia in Poland. Dr. Razzaghi has collaborated extensively with the Food and Drug Administration (FDA) as a Fellow of the National Center for Toxicological Research. In his role, he assisted in the mathematical development of dose-response models and statistical risk assessment procedures in animal bioassay experiments. He further studied methods for extrapolation of the procedures for human exposure to toxic chemicals in the environment. Dr. Razzaghi has also served as a consultant with the US Environmental Protection Agency (EPA) where he was a member of the peer review panel on the effects of perchlorate environmental contamination. Further, he has consulted with the National Institutes for Health (NIH) and Geisinger Medical Center. He has been the recipient of grants from the NIH and International Life Science Institute. Professional affiliations include membership with the American Statistical Association, Society for Risk Analysis, American Mathematical Society, and International Biometric Society, among others. He is also a Fellow of the Royal Statistical Society in England.

1

Introduction

1.1 Background

The science of toxicology has a long and profound history. It has its roots in the ancient Greek and Roman empires where physicians made early attempts to classify plants and distinguish between toxic and therapeutic plants. In the late fifteenth century and early sixteenth century, a Swiss scientist known as Paracelsus pioneered the use of chemicals and minerals in medicine. He is credited with the well-known phrase "The dose makes the poison." Although the original phrase was a little different, it emphasizes the fact that any substance can be harmful to living organisms if the dose or the concentration is high enough. Toxicologists believe that most chemicals, drugs, pollutants, and natural medicinal plants adhere to this principle. Paracelsus is often referred to as the father of toxicology. The development of modern toxicology is largely attributed to the Spanish scientist Orfila who is the father of modern toxicology and is considered to be the founder of the science of toxicology. Although different sources define toxicology a little differently, they all point to the fact that toxicology is the science that studies the adverse effects of chemical substances on living organisms including humans, animals, and the environment. It also involves the diagnosis and treatment of possible exposure to toxic substances and toxicants. Toxicologists use the power of science to predict how chemicals and damaging plants and minerals can be harmful. Not only is there individual variability in response, but other variables such as the exposure level, route of exposure, duration of exposure, age, gender, and the environment are used by toxicologists to determine the effect of toxicity.

1.2 Branches of Toxicology

Today, toxicology has grown into a multifaceted discipline. We provide here a list of these branches. The list is not exhaustive and there could be other branches as well, but it provides the most common branches with a brief description of each.

a. Aquatic Toxicology: Study of the effect of toxins such as chemical waste or natural material on aquatic organisms.

b. Chemical Toxicology: Involves the study of the structure and the mechanism of action of chemical toxicants.

c. Clinical Toxicology: Study concerning how much poison is present in the body as well as problems occurring due to overdose of drugs.

d. Developmental and Reproductive Toxicology: This branch of toxicology is concerned about the effect of toxins on the offspring when the parent, primarily the mother, is exposed to the toxicant during conception or pregnancy. It also concerns the multigenerational effects of toxic substances.

e. Ecotoxicology: Study of the effect of toxic substances in the ecosystem.

f. Environmental Toxicology: Study of the effects of pollutants naturally present in the environment, such as in air, water, and soil, on humans and other living organisms.

g. Forensic Toxicology: A topic within the general framework of forensic science that deals with the identification and quantification of poison, often leading to the determination of the cause of death.

h. Industrial Toxicology: Study of the effects of exposure to industrial waste and chemicals released from industries including, but not limited to, soot and other air pollutants.

i. Molecular Toxicology: This is concerned about the study of the cellular and molecular processes of toxicity.

j. Regulatory Toxicology: Study of toxicological processes based on the characteristics and guidelines of regulatory agencies.

k. Neurotoxicology: Study of the effects of toxicants on the brain and the nervous system.

l. Nutritional Toxicology: Concerned about food additives and nutritional habits, as well as the hazards posed by the way food is prepared, and so on.

m. Occupational Toxicology: Study of workplace-related health hazards, particularly in the chemical and mining industries.

n. Veterinary Toxicology: Study of the process of toxicity in animals.

o. Immunotoxicology: This is the study and analysis of how toxicity can damage the immune system.

p. Analytical Toxicology: Application of analytical chemistry methods in the quantitative and qualitative evaluation of toxic effects.

q. Mechanistic Toxicology: Similar to Chemical Toxicology, this study deals with the mechanism of action.

Some other branches of toxicology mentioned in the literature are Behavioral Toxicology, Comparative Toxicology, and Genetic Toxicology. Clearly many of these branches are interrelated and cannot be studied in isolation.

1.3 Basic Elements of Toxicology

Toxicity is defined as any undesirable or adverse effect of exogenous substances on humans, animals, and the environment. These substances include chemicals such as food additives, drugs and medicines, organic plants, or inorganic material such as mercury and lead. A specific undesirable outcome, such as carcinogenicity or neurotoxicity, is called a toxicological endpoint. Outcomes of toxicology testing experiments can be both continuous such as changes in brain weight and qualitative such as the presence or absence of a specific endpoint like cancer or can be evaluated on an ordinal scale such as low, moderate, or high. Toxicology tests have for a long time used laboratory animals in bioassay experiments to perform *in vivo* studies and to determine the effects of toxicants, although in recent years the use of *in silico* approaches (Computational Toxicology) has become popular. Experiments may consist of single exposure, as in case-control studies, or may have several exposure levels. The exposure level or the concentration level, that is, the amount of the chemical used in the experiment, is called the dose or the dosage level. Clearly, the dose is a crucial factor in the amount of toxicity, and determination of an efficient dose regimen is an important problem in the design of toxicological experiments. Several other factors play important roles in the extent of toxicity. One is the route of exposure, which could be by injection, oral (mixed in the diet), dermal, or by inhalation. Other factors are the frequency of exposure (how often the exposure occurs), duration of exposure, and the excretion rate of the chemical, often measured by half-life. Individuals respond differently to the same dosage, and other subject-specific variables such as age and gender add more variations to the outcome.

Toxicity is generally measured by the severity of the effect of the substance on the organism or the target tissue. The most fundamental method of measuring the toxicity of a substance is by using LD_{50}, which is the dosage level of the substance that creates lethality in 50% of the subjects. In inhalation toxicity studies, air concentrations are usually used for exposure values and LD_{50} is utilized as a measure of toxicity. Another similar measure is ED_{50} or EC_{50}, which is the effective dose or concentration of the chemical that makes an observable endpoint of interest in 50% of subjects. These measures have often been used to compare and classify chemicals. Clearly, 50% is a nominal and convenient value corresponding to the median, and other percentiles of interest may also be used. That is, in general, $LD_{100\,p}$ and $LC_{100\,p}$, where $0 \le p \le 1$ is the dosage or concentration level that results in lethality in 100%

of the subjects. Thus, for example, if $p = 0.01$, then ED_{01} refers to the dosage of the chemical that affects 1% of the subjects. Because humans are generally exposed to low levels of chemicals, much of the interest among toxicologists is to study the behavior and toxicity in the low-dose region. In fact, there was a large-scale experiment in the 1970s conducted by the National Center for Toxicological Research (NCTR) of the Food and Drug Administration (FDA) and reported by Staffa and Mehlman (1979), also referred to as the ED_{01} study. In that experiment, over 24,000 mice in several strains were exposed to the known carcinogen 2-acetylaminofluorene (2-AAF) to study the lethality of the chemical in low doses (see also Brown and Hoel, 1983a, b). However, LD_{50} and LC_{50} have limited usage as they cannot be directly extrapolated across species and to low doses. In fact, their application as a measure of toxicity has been criticized by many toxicologists (see Zbinden and Flury-Roversi, 1981; LeBeau, 1983). Alternative measures of toxicity are listed below:

a. Acceptable Daily Intake (ADI): For food additives and drugs.
b. Benchmark Dose (BMD): A dose of the toxin that produces a predetermined level (e.g. 5%) of change of the adverse effect.
c. Lowest-Observed-Effect-Level (LOEL): Lowest dose that causes an observable effect.
d. Lowest-Observed-Adverse-Effect-Level (LOAEL): Lowest dose that causes an observable adverse effect.
e. Maximum Tolerated Dose (MTD): Used mostly in chronic toxicology and represents highest dose with no health effects.
f. Median Tolerated Dose (TD_{50}): Median toxic dose causing toxicity in 50% of exposed individuals.
g. No Toxic Effect Level (NTEL): Largest dose with no observed effect.
h. No-Observed-Effect-Level (NOEL): Highest dose with no effect.
i. No-Observed-Adverse-Effect-Level (NOAEL): Largest experimental dose that produces no undesirable outcome.
j. Reference Dose (RfD): Daily acceptable dose that produces no risk of adverse effect.
k. Tolerable Daily (Weekly) Intakes: For contaminants and additives not consumed intentionally.
l. Reference Intake: Used mainly for nutrients.

There is a large body of literature in toxicology that describes the properties and applications of each of the abovementioned measures of toxicity. In addition, the measures are not independent and many of them are interrelated. Several publications discuss some of the relationships. For example, Gaylor and Gold (1995) and Razzaghi and Gaylor (1996) discuss the relation between TD_{50} and MTD.

1.4 Emergence of Statistical Models

Although statisticians have always played an important role in toxicological research and made contributions towards the development of many of the toxicological results, the earliest evidence of modeling applications can perhaps be attributed to Bliss (1934), who used the probit regression dose-response modeling and calculation of some toxicological measures. His research was concerned with controlling the insects that fed on grape leaves. He further developed the application of probit regression in Bliss (1935) and Bliss (1938). Later, Berkson (1944) applied logistic regression as an alternative to the probit model. The publication of the seminal book of Finney (1947), collecting the results of many of his earlier articles, contributed significantly to how statistical models can be used to make advances in the toxicological sciences, especially with respect to calculating and estimating risk. From that point on, in the 1970s and 1980s, a myriad of publications evolved that demonstrated the application of statistical models. At the same time, the development of many new statistical methodologies in various topics such as linear models (McCullagh and Nelder, 1989) and experimental designs (Box et al., 1978) and the demonstration of their application in biological sciences encouraged a number of collaborative works between biostatisticians and toxicologists. In this respect, the interests and promotion of research by regulatory agencies such as the FDA, the National Institutes of Health (NIH), and the Environmental Protection Agency (EPA) are noteworthy. The establishment of research centers in these branches of the federal government, such as the NCTR by the FDA, the National Cancer Institute by the NIH, and the National Institute of Environmental Health Sciences by the EPA played a major role. The fundamental contributions of their statisticians were highly instrumental in the development of the application of statistical models and their extensions in toxicology. Their research and results were not only important and influential in the use of mathematical models, but more importantly encouraged numerous collaborative works between the government and researchers in academic institutions as well as private enterprises and industries, leading to a large number of publications with pivotal results. The seminal book of Collette (1991) is particularly noteworthy. Moreover, the advancement of computer technology and the ability to perform complex calculations further contributed to the creation of *in silico* toxicology, which is a type of toxicity assessment that uses computer models to simulate and visualize chemical toxicities. According to Raies and Bajic (2016), the modeling method is an important aspect of *in silico* toxicology, and steps to generate a prediction model include model generation, evaluation, and interpretation.

A publication of the National Research Council (NRC, 1983) points out that biostatistical models, particularly quantitative evaluation, of toxicity are crucial in toxicological research and can help in four ways:

1. Developing an overall testing strategy to set dose and exposure regimens.
2. Optimal design to extract maximum information.
3. Interpretation of data.
4. Verifying the underlying biological assumptions.

Today, regulatory agencies and pharmaceutical industries rely heavily on the biostatistical models that provide estimates of health risk to humans and other organisms, and research is ongoing to explore models that improve the accuracy of such estimates.

1.5 Scope of This Book

This book is about statistical models. An attempt is made to present an account of the most commonly used mathematical and statistical models in toxicology. Clearly, not every branch of toxicology is covered. Since the early 1970s, many statistical models have been developed for expressing toxicological processes mathematically. Hoel (2018) refers to this period as an "exciting time" because of the attention given by statisticians to the problem of estimating the human health risk of environmental and occupational exposures. Not only have the models enjoyed a high level of elegance and sophistication mathematically, they were widely used (and for the most part are still being used) by industry and government regulatory agencies. Thus, in this book we are primarily concerned with the models developed for the assessment of human health risk. Toxicologists have used animal models and bioassay experiments for a long time to understand the mechanism of toxicity in order to be able to estimate the risk of exposure. Therefore, the focus of this book is to describe the statistical models in environmental toxicology that facilitate the assessment of risk mainly in humans. For this purpose, the basic concepts and methods of risk assessment are described in Chapter 2. Since humans are rarely exposed to a single chemical in isolation and exposure is often to a mixture of chemicals, models that assess the risk of mixtures of chemicals are discussed in Chapter 3. In Chapters 4 through 7, we present statistical models that are developed for risk estimation in different aspects of environmental toxicology. The problem of modeling and risk analysis for cancer and exposure to carcinogenic substances is presented in Chapter 4. Note that the outcomes of carcinogenicity experiments are binary in nature since the experimental unit either contracts cancer or is free of cancer. In toxicology, methods for risk assessment of cancer endpoints differ from those of non-cancer endpoints. In Chapter 5, we discuss models for developmental and reproductive toxicity risk assessment. These experiments are designed and conducted to assess the health risk for the offspring when the mother is exposed to a toxin during pregnancy. In these experiments, toxicologists observe whether or not the offspring suffers from any abnormality such as malformation or death. Consequently, except in cases where fetal weight is under consideration, the outcomes from developmental toxicity experiments are, for the most part, dichotomous in nature. Chapter 6 is devoted to describing the statistical models for risk assessment in continuous outcomes. Specifically in experiments designed to assess the effect of toxic substances on the brain and the nervous system, the outcomes are continuous in nature. Finally, in Chapter 7, we present statistical models for developmental neurotoxicity. These models are developed for assessing the risk and possible effect on the nervous system of the offspring when the mother is exposed to harmful substances during pregnancy.

It should be emphasized that with the development of modern technology and the advent of the digital revolution, the protocol for some of the experiments designed for toxicity testing has been changed or modified. Thus, some of the models described in this book may no longer be in use several years from now. But, the fact is that the mathematical elegance of these models alone makes them quite interesting to study. Moreover, statistical models for more recent protocols have still not been fully developed, and research in those areas is ongoing.

References

Berkson, J. (1944). Application of the logistic function to bioassay. *Journal of the American Statistical Association*, 39, 357–65.

Bliss, C. I. (1934). The method of probit analysis. *Science*, 79, 38–9.

Bliss, C. I. (1935). The calculation of dosage mortality curve. *Annals of Applied Biology*, 22, 134–67.

Bliss, C. I. (1938). The determination of dosage mortality curve from small numbers. *Quarterly Journal of Pharmacology*, 11, 192–216.

Box, G. E. P., Hunter, W. G., and Hunter, J. S. (1978). *Statistics for experimenters: An Introduction to design, data analysis, and model building*. Wiley, New York.

Brown, K. G. and Hoel, D. G. (1983a). Multistage prediction of cancer in serially dosed animals with application to the ED_{01} study. *Toxicological Sciences*, 3, 470–77.

Brown, K. G. and Hoel, D. G. (1983b). Modeling time to tumor data: Analysis of ED_{01} study. *Toxicological Sciences*, 3, 458–69.

Collette, D. (1991). *Modeling binary data*. Chapman and Hall, London.

Finney, D. J. (1947). *Probit analysis*, 1st edition. Cambridge University Press, Cambridge.

Gaylor, D. W. and Gold, L. S. (1995). Quick estimate of the regulatory virtually safe dose based on the maximum tolerated dose for rodent bioassay. *Regulatory Toxicology and Pharmacology*, 22, 57–63.

Hoel, D. G. (2018). Quantitative risk assessment in the 1970s: A personal remembrance. *Dose-Response*, 16, 1559325818803230. PMID: 30302069.

LeBeau, J. E. (1983). The role of the LD_{50} determination in drug safety evaluation. *Regulatory Toxicology and Pharmacology,* 3, 71–4.

McCullagh, P. and Nelder, J. (1989). *Generalized linear models.* Chapman and Hall, London.

National Research Council (1983). *Risk assessment in the federal government: Managing the process.* National Academy Press, Washington, DC.

Raies, A. B. and Bajic, V. B. (2016). In silico toxicology: Computation methods for the prediction of chemical toxicity. *Wiley Interdisciplinary Reviews: Computational Molecular Sciences,* 6, 147–72.

Razzaghi, M. and Gaylor, D. G. (1996). On the correlation coefficient between the TD50 and the MTD. *Risk Analysis,* 16, 107–13.

Staffa, J. A. and Mehlman, M. A. (1979). Innovations in cancer risk assessment (ED01 study). In Proceedings of a Symposium Sponsored by the National Center for Toxicological Research and the US Food and Drug Administration. J. A. Staffa and M. A. Mehlman eds. Pathotox Publishers, Inc.

Zbinden, G. and Flury-Roversi, M. (1981). Significance of the LD50-test for the toxicological evaluation of chemical substances. *Archives of Toxicology,* 49, 99–103.

2

Quantitative Risk Assessment

2.1 Introduction

Risk assessment is the process of evaluating and deriving a quantitative or qualitative estimate of the potential health risk to humans or other organisms from a defined source of hazard. It represents the likelihood of an adverse effect or the absence of a beneficial effect. The task force on health risk assessment (USDHHS, 1985) distinguishes between human health risk assessment and ecological risk assessment. Accordingly, human health risk assessment is defined as the characterization of the probability of potential adverse health effects as the result of human exposure to chemical hazards through various sources such as medicine, food additives, and chemicals present naturally in the environment. Ecological risk assessment, on the other hand, is the process of estimating the probability of undesirable ecological effects occurring due to human activities. Risk is generally defined as the probability of an adverse effect, and risk estimation can potentially present several problems due to scarcity and uncertainty in the data. In this chapter, we first discuss the process of risk assessment formalized by the National Research Council. We then describe some common approaches to quantitative risk assessment. Attention is focused on the critical stage of dose-response evaluation where statistical models play a key role in the quantification of risk, and the crucial choice of the model has a fundamental role in risk estimation.

2.2 Process of Quantitative Risk Assessment

Gathering information and evaluating that information in order to determine the quantitative value of the risk associated with exposure defines the process of risk assessment. The National Research Council in a publication (NRC, 1983) that later became known as the "Red Book" formalized this process. In this publication, the process of risk assessment was defined as consisting of four distinct and non-overlapping stages: Hazard Identification, Exposure

Assessment, Dose-Response Assessment, and Risk Characterization. Below is a brief description of each of these four stages:

a. Hazard Identification: This stage consists of the qualitative identification of sources of hazard—whether the hazard source was due to a chemical substance or conditions under which the injury, illness, or disease occurred. It also entails identification of the types and extent of adverse effects as well as the substance that may cause potential health hazard.

b. Exposure Assessment: This is the stage where possible routes of exposure are identified and a quantitative evaluation is made of the extent of exposure. The questions as to who was exposed, at what level exposure occurred, and for how long are posed and studied in this stage. Quite often answers to these questions may not be readily available and this stage can be quite difficult to evaluate. The level and duration of exposure could be hard to determine especially when exposure occurs irregularly at different intervals of time and at varying levels.

c. Dose-Response Assessment: Perhaps the most crucial stage, it involves the quantification of the relationship that exists between the risk and the exposure level. A sigmoid-shaped mathematical expression describing the effect as a function of dose is often used to show this relationship. The choice of this function plays an extremely important role in risk estimation. In fact, it can be shown that the risk estimate can vary by orders of magnitude using two very plausible functions. Different sources of information such as chemical, epidemiological, and toxicological data are used to determine this mathematical relationship. It is often argued that a dose-response relationship which is built based on the underlying biological processes would have a much better chance in estimating the risk more accurately. A principal goal of establishing the dose-response relationship is to determine an exposure level of the toxin that may be used as the so-called *Point of Departure* (POD) or the starting point for calculating an acceptable exposure levels for the human population. It is defined as the point on the dose-response curve corresponding to a low fixed nominal effect level. Clearly that can also depend on the group of target individuals and one should take into consideration the susceptibility of the group. For example, the characterization of the POD may be different for children than the adult populations. In later chapters of this book, we pay special attention to this stage of risk assessment and focus on the construction of dose-response models in various branches of toxicology.

d. Risk Characterization: This is the last and the final stage of risk assessment. It involves the integration of the information gathered from the previous three stages into a quantitative determination of the risk. It often includes a discussion of the uncertainties associated with the estimate of risk.

It should be emphasized that we differentiate between risk assessment and risk management. Although the goal of both procedures is the characterization of risk, the approach and the type of information used in risk assessment are different. Torres and Bobst (2015) provide a detailed step-by-step account of the above stages of risk assessment.

2.3 Methods of Risk Assessment

In toxicology, in order to assess the adverse effects of chemical compounds, animal bioassay experiments are designed and the results are utilized within the framework of risk assessment to estimate the risk of exposure. Laboratory animals, mostly mice and rats and other rodents, are exposed in controlled environments, and the effects are measured. There are several statistical problems to consider even at the design stage of toxicological experiments. For example, the dosage regimen, route of exposure, and duration of exposure are all questions that must be answered for designing the experiment. Generally, in order to elicit the chemical effects in a limited number of animals, dosage levels much higher than human exposure levels are used. Therefore, one of the fundamental problems in toxicology is the extrapolation from high experimental doses to low environmental dosage levels. In addition, extrapolation from animal doses to human exposure levels is another problem that toxicologists encounter. There are also other problems that are addressed and considered such as varying animal susceptibility. As described before, the goal of quantitative risk assessment is to estimate the probability of an adverse effect on humans or other organisms as the result of exposure to a very low dosage level. Equivalently, the problem of risk assessment can be posed as the determination of a safe exposure level for a small negligible risk. Regulatory agencies and also the pharmaceutical industry rely on these estimates for setting allowable daily intakes for humans.

The method of risk assessment in toxicology depends on the endpoint. There are generally two approaches taken depending on whether the endpoint is cancer or non-cancer. Moreover, the approaches for quantitative derivation of risk depend on whether or not the response of interest is quantal or quantitative.

a. **Risk Assessment for Cancer Effects**

In toxicology, estimation of cancer risk is usually carried out on the basis of long-term bioassay experiments. In such experiments, test animals are divided into several groups, usually with equal numbers of animals. Each group is then exposed to a different level of the chemical including a control group of untreated animals. Exposure generally occurs for a period of 200 days to two years. The methodology for cancer risk assessment is somewhat different as it is generally believed that there is no threshold in cancer dose-response models, and even very low levels of exposure to a carcinogenic substance may increase the risk of developing cancer. Even if a threshold does exist, it is considered to be very low that it cannot be reliably identified. For carcinogenic substances, therefore, the relationship between cancer incidence and the dose of the chemical observed in an experimental study is extrapolated to the lower doses in such a way that an excess lifetime risk of cancer—that is, the added risk of cancer resulting from lifetime exposure to that chemical at a particular dose is a fixed negligible value. Thus, while there is no "safe" dose with a risk of zero, at sufficiently low doses the risk becomes very low and is regarded to have minimal to no adverse health effects. Cancer risk is generally assumed to be proportional to dose at low exposure levels. Although this conjecture is difficult to prove, there are many experimental evidences that support the assumption. In practice therefore, a dose-response model is fitted to the experimental data and a statistical upper confidence limit is established at the lowest experimental dose that causes cancer. This value is called the *cancer slope factor* and it is assumed to be the upper bound estimate of the probability that an individual will develop cancer due to a lifetime exposure to the chemical agent. A linear extrapolation with no threshold is then used to estimate the effect for lower doses. Thus, the line connecting the cancer slope factor to the origin is used to estimate risk at low exposure levels. Figure 2.1 is an illustration of this extrapolation process. Similarly, to estimate safe exposure levels, a lower confidence limit on the dose corresponding to a small nominal risk (such as 10%) is determined and is used as the POD. Linear extrapolation with no threshold is utilized to find safe exposure levels below that level. A wide variety of dose-response models are used for modeling cancer risk, and in Chapter 4 we discuss many of these models. However, as will be pointed out later, although several models may provide adequate fits to the data in the experimental dose levels, when extrapolated to low dosage levels, the estimate of the "safe" exposure level that corresponds to a very low negligible risk may vary quite substantially by orders of magnitude. Thus, the EPA guidelines (US EPA, 2005) does not rely on recommending a specific model for the purpose of risk assessment. Rather, it is suggested that to determine a POD, several dose-response models may be considered for estimating a statistical lower confidence bound for a dose corresponding to a low risk of say 1% to 10%.

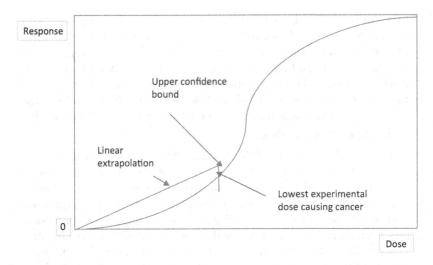

FIGURE 2.1
Low-dose extrapolations for cancer effect.

b. Risk Assessment for Non-Cancer Effects

For non-cancer effects, the approach for risk assessment is based on the assumption of the existence of a threshold. Thus, it is assumed that there is an exposure level below which there is no biologically or statistically significant adverse effect. Clearly, the estimation of the exact threshold dose from experimental data is not trivial. The traditional method as summarized by the US Environmental Protection Agency (US EPA, 2005), is to first establish a No-Observed-Adverse-Effect-Level (NOAEL) from an animal bioassay experiment. This is the highest dose level which produces no statistically or biologically significant effect compared to the background or the control animals. The NOAEL is used as the POD defined as the dose-response point that marks the beginning of low-dose extrapolation. To determine a safe exposure level, the NOAEL is divided by a series of safety factors (SFs) or uncertainty factors (UFs) to account for uncertainties due to animal-to-human extrapolation, high-to-low-dose extrapolation, and other possible uncertainties to determine the so-called Reference Dose (RfD). The RfD is defined as an estimate of a daily exposure level that is unlikely to cause any appreciable risk of adverse effect to humans. Thus,

$$RfD = \frac{NOAEL}{UF_1 \times UF_2 \times \cdots \times UF_k}$$

When NOAEL is not available, a LOAEL is used. The NOAEL approach has been sharply criticized by many researchers as it is ill defined, rather arbitrary, and does not reward better experimentation.

In addition, NOAEL method depends on the dosing regimen and the dose selection. Moreover, determination of NOAEL does not utilize the shape of the dose-response curve and also depends on sample size. Due to these limitations, the NOAEL/LOAEL approach has been criticized by many authors, see for example Leisenring and Ryan (1992), Allen et al. (1994), and Barnes et al. (1995), and model-based approaches have instead gained more widespread popularity.

c. **Benchmark Dose Methods**

The benchmark dose (BMD) methodology was introduced by Crump (1984) as an alternative approach in response to the short-falls of the NOAEL approach for risk estimation for non-cancer endpoints. It is a powerful methodology that has found widespread popularity since its inception. The main advantage of the methodology is that it utilizes the shape of the dose-response relationship and it is not limited to only the experimental dose levels. The method involves fitting a statistical model to the dose-response data and calculating the BMD as the dose that causes a fixed preset change in response, called the benchmark response BMR (usually 5% or 10%). The statistical lower confidence (usually 95%) limit (BMDL) is then determined and used as the POD. Thus, rather than dividing the NOAEL by the uncertainty factors, it was proposed to utilize the BMDL. After its introduction by Crump (1984), many properties of the BMD methodology were discussed by several authors. We refer to the most recent comprehensive review of the subject by Haber et al. (2018) and references therein. The main advantages of using BMDL instead of NOAEL as POD can be listed as follows:

- Accounts for the shape of the dose-response curve and hence it is not limited to the experimental doses.
- The dose spacing is less critical.
- Rewards better experimentation as lower variability leads to more accurate BMDL.
- Larger sample sizes result in lower uncertainty in the calculation of BMDL.
- It is more consistent and allows for comparison across chemicals.

The BMD methodology was introduced primarily to replace the NOAEL approach for finding the POD in non-cancer risk assessment and the early discussion centered mostly around the quantal responses from developmental toxicity studies. However, because of its many interesting properties, it was soon realized that the methodology can effectively also be used in cancer risk assessment. Gaylor, et al. (1999) argue that the basis and impact of the dichotomy between the procedures for risk assessment in cancer and non-cancer endpoints are questionable. They propose using the BMD methodology for determining the POD, and dividing the BMDL by the uncertainty factors. In fact, the carcinogenic risk assessment guidelines by the US EPA (1996) selected the benchmark dose approach for low-dose cancer risk assessment. Later, in 2009, the Committee on Improving Risk Analysis Approaches Used by the US EPA, charged by the National Research Council, conducted a comprehensive review of approaches utilized for risk assessment and provided recommendations for practical improvements of the approaches. In their report, NRC (2009), the dichotomous approach is criticized and a unified approach that "develops an integrative framework that provides conceptual and methodologic approaches for cancer and non-cancer assessments" based on the BMD methodology is clearly recommended. Further, the EPA (2012) document on benchmark dose clearly stresses "harmonization of approaches for cancer and noncancer risk assessment." For a recent review of the BMD methodology, we refer to Haber et al. (2018).

An issue in the characterization of dose-response assessment is the choice of the right model since several models may fit the data well statistically, but when extrapolated to low doses they can produce different estimates of the safe exposure level. Although the use of BMDL as the POD helps in reducing the ambiguity in the choice of model, to further reduce uncertainty and derive more accurate estimates, in recent years an approach based on model averaging where a weighted average of several models is used has found widespread popularity. A Bayesian method is used to determine the weights in such a way that models with a better fit have relatively higher weight. We discuss model averaging technique and its application in developmental toxicity risk assessment in Chapter 5. Now, since the process of risk assessment using the BMD methodology varies slightly depending on whether or not the outcome of interest in the experiment is quantal or quantitative, we discuss these methods for the two types of outcomes. But, first we define the quantitative measures of risk.

2.4 Quantitative Definitions of Risk

In toxicology, risk is generally defined as the measure that an adverse effect will occur under defined conditions of exposure to a chemical. More formally, let $\pi(d)$ denote the risk function, i.e. the probability that an adverse effect is observed as the result of exposure to a concentration level d of a toxic substance. Then, there are commonly two mathematical definitions of the measure of risk:

a. **Additional Risk:** This is the difference in risk or the probability of response in the exposed and unexposed subjects. Thus, it is a measure that provides the increase in the probability of adverse response for individuals exposed to a certain dose d as compared to the background. Therefore,

$$\pi_A(d) = P(d) - P(0) \tag{2.1}$$

The additional risk is also called the **Excess Risk**.

b. **Extra Risk**: This is defined as the ratio of the additional risk to the probability of observing an effect in non-exposed individuals. Thus, it is the proportional increase in risk adjusted for the background. Therefore,

$$\pi_R(d) = \frac{P(d) - P(0)}{1 - P(0)} \tag{2.2}$$

The extra risk is also referred to as the **relative risk.**

Note that both measures are between 0 and 1 and that

$$\lim_{d \to \infty} \pi_R(d) = 1 \tag{2.3}$$

2.5 Risk Assessment for Quantal Responses

The outcome of interest in many toxicological studies is qualitative in nature. In a vast majority of these experiments, the outcome is quantal and binary. For example, in cancer studies, animals are exposed to a dosage of a carcinogenic chemical for a certain period of time and at the conclusion of the experiment animals are examined for presence or absence of tumor. Similarly, in developmental toxicity studies, pregnant female animals are exposed to a known teratogen during a certain critical time of pregnancy. The animals are sacrificed just before term, and the fetuses are examined for presence or absence of a malformation. Sometimes in these experiments multiple outcomes may be of interest and we will discuss the statistical models for developmental toxicity studies in Chapter 5. For experiments with dichotomous responses, several statistical models with sigmoid shape graphs have been introduced. Table 2.1 gives a list of the most popular dose-response models for quantal responses where d is the exposure level and $P(d)$ is the probability of response.

The process of risk assessment using the benchmark dose methodology for quantal responses begins by first fitting a dose-response model to the data. A value of BMR, generally a value between 1% and 10% is then selected as the added risk. Depending on whether (2.1) or (2.2) is used as the definition

TABLE 2.1

Common Dose-Response Models for Quantal Response

Title	Mathematical Formula
One-Hit (Exponential)	$P(d) = 1 - \exp\left[-(\beta_0 + \beta_1 d)\right] \quad \beta_j \geq 0; \; j = 0,1$
Logistic	$P(d) = \left[1 + \exp-(\beta_0 + \beta_1 d)\right]^{-1} \quad -\infty < \beta_0 \langle \infty, \beta_1 \rangle 0$
Probit[a]	$P(d) = \Phi(\beta_0 + \beta_1 d) \quad -\infty < \beta_0 \langle \infty, \beta_1 \rangle 0$
Log-Logistic	$P(d) = c + (1-c)\left[1 + \exp-(\beta_0 + \beta_1 \log d)\right]^{-1} \quad 0 \leq c < 1, \; -\infty < \beta_0 \langle \infty, \beta_1 \rangle 0$
Log-Probit	$P(d) = c + (1-c)\Phi(\beta_0 + \beta_1 \log d) \quad 0 \leq c < 1, \; -\infty < \beta_0 \langle \infty, \beta_1 \rangle 0$
Weibull	$P(d) = 1 - \exp\left[-(\beta_0 + \beta_1 d^\gamma)\right] \quad \beta_j \geq 0; \; j = 0,1, \; \gamma > 0$
Multi-Hit (Gamma)[b]	$P(d) = c + (1-c)\left[\Gamma(\beta_0)\right]^{-1} \int_0^{\beta_1 d} t^{(\beta_0 - 1)} e^{-t} dt \quad 0 \leq c < 1$ $\beta_0 \geq 0, \; \beta_1 \geq 0$
Multistage	$P(d) = 1 - \exp\left[-\sum_{j=0}^{k} \beta_j d^j\right], \; \beta_j \geq 0; \; j = 0,1,\ldots,k$
Dichotomous Hill Model[c]	$P(d) = \theta \dfrac{1 + \gamma \exp\left[-(\beta_0 + \beta_1 \log d)\right]}{1 + \exp\left[-(\beta_0 + \beta_1 \log d)\right]}, \; 0 < \theta \leq 1, 0 \leq \gamma < 1, \; \beta_j \geq 0; \; j = 0,1$

Note: c denotes the background response.
Notes: [a]$\Phi(.)$ is the cdf of the standard normal distribution. [b]$(.)$ is the gamma function. [c]θ is the maximum probability of response predicted by the model.

of added risk, the benchmark dose can be determined. A lower confidence limit (BMDL) is used as the POD and linearly extrapolated to the origin. Figure 2.2 gives an illustration of this methodology. The mathematical criteria for computing the lower confidence limit on BMD will be discussed in later chapters. Although the two definitions can lead to different values, for small doses that difference is often negligible. In Chapters 4 and 5 we apply the risk assessment methods for binary responses and explore the applications of several models listed in Table 2.1.

2.6 Risk Assessment for Quantitative Responses

The process of risk assessment when the response of interest in the experiment is quantitative begins by first fitting an appropriate dose-response model. This is a mathematical relationship that expresses the mean response $\mu(d)$ as a function of the administered dose d. Once again, there is a wide

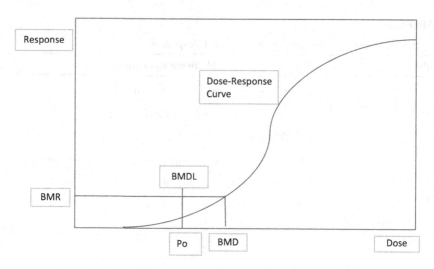

FIGURE 2.2
Illustration of the benchmark dose methodology.

TABLE 2.2

Common Functions for Mean Response in Quantitative Responses

Title	Mathematical Formula
Polynomial	$\mu(d) = \beta_0 + \beta_1 d + \cdots + \beta_k d^k$
Power	$\mu(d) = \beta_0 + \beta_k\, d^k$
Exponential	$\mu(d) = \beta_0 + \beta_1\left(1 - \exp\left[-\left(\dfrac{d}{\tau}\right)^\gamma\right]\right)$
Hill	$\mu(d) = \beta_0 + \beta_1 \dfrac{d^k}{\tau^k + d^k}$
Michaelis-Menten	$\mu(d) = \beta_0 + \beta_1 \dfrac{d}{\tau + d}$

range of models that have been introduced for continuous responses and Table 2.2 provides a list of the most common statistical models. But, when the response is quantitative, a fundamental question that needs to be answered before applying a risk assessment method is what response constitutes an adverse effect. In the absence of a clear definition of an undesirable response, in order to define the BMR and consequently the BMD, a few approaches have been proposed. Crump (1984) defined the BMD as the dose corresponding to a response that causes a certain percentage change in response relative to the background response $\mu(0)$. Depending on whether low or high responses are considered undesirable, this can be expressed as

$$\mu(\text{BMD}) = \mu(0) \pm \text{BMR}\mu(0) \qquad (2.4)$$

Slob and Pieters (1998) discussed a response that causes some level of change relative to the background and used the term Critical Effect Size (CES) for this response. Alternatively, BMD may be defined as the exposure level that causes a change in mean response relative to the standard deviation of the background responses $\sigma(0)$, that is

$$\mu(\text{BMD}) = \mu(0) \pm \text{BMR}\sigma(0) \tag{2.5}$$

Kodell and West (1993) defined an adverse effect for as a response that is rare in control animals and falls more than a fixed number k standard deviation from the mean response of the control distribution. For example, when the responses have a normal distribution, k could be 2.33 or 3. Therefore, one can define the BMD as the dose level that produces a response below a certain value in the background distribution. This method is called the hybrid approach and was used by Crump (2002) and West and Kodell (1999). We will consider this issue and the role that choice of k plays in risk assessment in length in Chapters 6 and 7 when we discuss continuous responses from neurotoxicity and developmental neurotoxicity experiments.

2.7 Multivariate Responses

In toxicological bioassay experiments, especially in non-cancer studies, seldom a single measurement characterizes the toxic response to the chemical compound, and often several measurements are taken on each experimental unit to show the overall toxicity of a substance. For example, in developmental toxicity studies, although the presence or absence of a malformation may be of primary interest, other measurements such as the fetal weight and the location of the fetus in the litter may be of interest. Characterization of the BMD methods for multiple responses has not yet fully been developed and currently BMD is calculated for each response although a joint BMD may be preferred. For the case of bivariate quantitative responses, Regan and Catalano (1999) consider separate cut-off values for the two responses and define the probability of an adverse response using a bivariate dose-response model. Yu and Catalano (2005) applied the bivariate normal distribution as the dose-response model to derive joint BMDs for paired responses from a neurotoxicity study triethyl exposure in rats.

2.8 Benchmark Dose Software

With the growth of popularity and importance of the BMD methodology in risk assessment, the US EPA developed the benchmark dose software in 1999 to facilitate the process of model fitting and calculation of the RfD and hence determination of the POD. The development has gone through many updates and the most current version, BMDS (2017) contains several modifications and is used by thousands of users and risk assessors worldwide. The software is very versatile and available for download on the EPA web site. It is an MS EXEL–based program that is highly user friendly and simple to use. The software can be used for cancer quantal data with single or multiple tumors as well as for continuous outcomes. In addition, there is an option for nested data from developmental toxicity experiments. Several choices of dose-response models for each type of data are available with the choice of additional or extra risk as defined in (2.1) and (2.2) as measures of risk. To alleviate the burden of choosing a single dose-response model, the software provides the option of using the Bayesian Model Averaging technique with user provided weights. The output provides information about the model parameter estimates, goodness-of-fit, BMD, and BMDL for each model. For quantitative responses options are available for choosing the adverse effect based on the number of standard deviation or the hybrid model. For more information about the software, refer to the EPA web site https://www.epa.gov/bmds.

Example 2.1

To illustrate the BMD procedure and the BMDS, we use the dose-response data from an experiment that studies the effect of trichloro-ethylene (TCE) in rats. The chemical is a nonflammable, colorless liquid used mainly as a solvent for removing grease from metal parts, and also as an ingredient in adhesives and paint removers. TCE is known to be carcinogenic to humans. Maltoni et al. (1986) report the results of a large-scale study consisting of eight experiments involving the exposure of nearly 4000 animals to the chemical TCE. For illustrative purposes, we use the results of the incidence of leukemias in Sprague-Dawley male rats exposed to four doses of TCE by inhalation for 7 hours daily, 5 days per week, and for a duration of 104 weeks. For more information about the experiment we refer to Maltoni et al. (1986) and the toxicological review of TCE published by EPA (2011). Table 2.3 provides information on the four dosage levels used in the experiment along with the number of rats alive at the end of the experiment and the number affected with leukemia. The multistage model was selected for dose-response modeling. The multistage model has been the recommended choice for cancer bioassay data by the US EPA for a long time (Figure 2.3). If the multistage model is selected in the BMDS, the program automatically fits all models up to the number of dose levels less 1. Thus, in the current data set, since there are four dose levels, the program fits multistage models of up to degree $k=3$ (i.e. linear, quadratic, and cubic). In addition, the program has a choice of restricting the model parameters to be non-negative. Since, the US EPA recommend this restriction, it was selected for the current analysis. In addition, extra risk was selected as the measure of risk and an added risk of 0.1 was used as the benchmark risk (BMR). Table 2.4 provides the summary information for the model fitting from the BMDS

TABLE 2.3

Incidence of Leukemia in Rats Exposed to TCE

Dose (ppm)	Number Alive	Number with Cancer	Proportion with Cancer
0	134	9	0.067
100	130	13	0.1
300	130	14	0.108
600	129	15	0.116

Source: Matoni et al. (1986), Table 17, P. 319, EPA (2011).

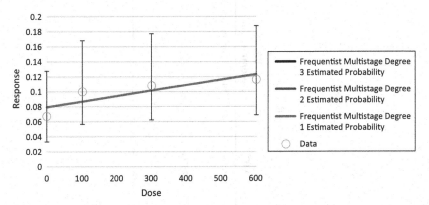

FIGURE 2.3
Model summary with BMR of 10% extra risk for the BMD and 0.95 lower confidence limit for the BMDL.

output. The software recommends the use of the multistage model with the third degree due to its lowest value of the AIC although from the output it is clear that there is not much difference between the three models in this case. The value of the BMDL which can be used as the POD for linear extrapolation is 533.4925 for the selected dose-response model.

Exercises

1. Suppose that the dose-response relationship in an animal bioassay experiment with a binary response is characterized by the Weibull model

$$P(d) = 1 - \exp\left[-\left(\beta_0 + \beta_1 d^\gamma\right)\right]$$

where $P(d)$ is the probability of response at dose d and β_0, β_1, and γ are model parameters.

TABLE 2.4

Summary Information for Dose-Response Modeling of the Incidence of Leukemia in Rats

Model	BMD	BMDL	P Value	AIC	BMDS Recommendation	BMDS Recommendation Notes
Multistage degree 3	1293.168	533.4925	0.715618	336.7539812	Viable – recommended	Lowest AIC
Multistage degree 2	1293.161	533.498	0.715618	336.7539812	Viable – alternate	BMD higher than maximum dose
Multistage degree 1	1293.174	533.5041	0.715618	336.7539812	Viable – alternate	BMD higher than maximum dose

a. Show that $P(d)$ can be expressed as

$$P(d) = 1 - (1-c)e^{-\beta_1 d^\gamma}$$

where $c = P(0)$ is the background response rate.

b. Derive the expressions for the additional risk $\pi_A(d)$ and the relative risk $\pi_R(d)$.

c. Show that for small doses risk is a linear function of dose if and only if $\gamma = 1$.

2. Table 2.5 gives the data from Maltoni et al. (1986) for the incidence of tetis, Leydig cell tumor in rats exposed to TCE (see also US EPA, 2011). Use the BMDS software for dichotomous response with the restricted multistage model as the dose response to determine the BMD, BMDL, and hence the POD. Experiment with other choices of dose-response models and provide a report on the effect of the choice of model on POD.

3. The toxicity of the dose-dependent response for the inhibition of the tumor growth for 1α-hydroxyvitamin D_2 (1α-OH-D_2) in mice was reported by Grostern et al. (2002). A total of 292 athymic mice were injected with the chemical at four non-zero doses plus the control. Table 2.6 gives the summary data on mean and standard

TABLE 2.5

Incidence of Tetis, Leydig Cell Tumor in Rats Exposed to TCE

Dose (ppm)	Number Alive	Number with Cancer	Proportion with Cancer
0	121	6	0.05
100	116	16	0.138
300	116	30	0.259
600	122	31	0.254

Source: Matoni et al. (1986).

TABLE 2.6

Dose-Response Data for Mice Exposed to 1α-OH-D_2

Dose (μg)	Body Weight (g) Mean	SD	Tumor Weight (g) Mean	SD	Tumor Size (mm²) Mean	SD
Control	18.43	0.64	1.76	0.51	2914.88	713.03
0.1	18.58	0.86	1.33	0.56	1688.09	594.23
0.2	17.01	0.95	0.63	0.3	1087.23	436.18
0.3	15.46	0.69	0.73	0.23	1102.09	291.18
0.6	15.81	0.71	0.96	0.31	1251.49	344.38

Source: Data from Grostern et al. (2002).

deviation for three categories of the body weight, tumor weight, and the tumor size for dose group. Use the BMDS software for continuous responses to estimate the BMDs and BMDLs for each category of response. Assume a constant variance and a normal distribution of the mean responses and use the hybrid approach to define an adverse effect. Experiment with different dose-response models and provide a report of your findings.

References

Allen, B. C., Kavlock, R. J., Kimmel, C. A., and Faustman, E. M. (1994). Dose-response assessment for developmental toxicity II. Comparison of genetic benchmark dose estimates with NOAELs. *Fundamental and Applied Toxicology*, 23, 487–495.

Barnes, D. G., Daston, G. P., Evans, J. S., Jarabek, A. M., Kavlock, R. J., Kimmel, C. A., Park, C., and Spitzer, H. L. (1995). Benchmark dose workshop: Criteria for use of a benchmark dose to estimate a reference dose. *Regulatory Toxicology and Pharmacology*, 21(2), 296–306.

BMDS (2017). Benchmark Dose Software – Version 3.0. US Environmental Protection Agency.

Crump, K. S. (1984). A new method for determining allowable daily intakes. *Fundamental of Applied Toxicology*, 4, 854–871.

Crump, K. S. (2002). Critical issues in benchmark dose calculations from continuous data. *Critical Review in Toxicology*, 32, 133–153.

Gaylor, D. G., Kodell, R. L., Chen, J. J., and Krewski, D. (1999). A unified approach to risk assessment for cancer and noncancer endpoints based on benchmark doses and uncertainty/safety factors. *Regulatory Toxicology and Pharmacology*, 29, 151–157.

Grostern, R. J., Bryar, P. J., Zimbric, M. I., Darjatmoko, S. R., Lissauer, B. J., Lindstrom, M. J., Lokken, J. M., Strugnell, S. A., and Albert, D. M. (2002). Toxicity and dose-response studies of 1α-Hydroxyvitamin D_2 in a retinoblastoma xenograft model. *Archives of Ophthalmology*, 120, 607–612.

Haber, L. T., Dourson, M. L., Allen, B. C., Hertzberg, R.C., Parker, A., Vincent, M. J., Maier, A., and Boobis, A. R. (2018). Benchmark dose (BMD) modeling: Current practice, issues, and challenges. *Critical Review in Toxicology*, 48, 387–415.

Kodell, R. L. and West, R. W. (1993). Upper confidence limits on excess risk for quantitative responses. *Risk Analysis*, 13, 177–182.

Leisenring, W. and Ryan, L. (1992). Statistical properties of the NOAEL. *Regulatory Toxicology and Pharmacology*, 15, 161–171.

Maltoni, C., Lefemine, G., and Cotti, G. (1986). Experimental research on trichloroethylene carcinogenesis. In *Archives of research on industrial carcinogenesis* volume 5, C. Maltoni and M. A. Mehlman eds. Princeton Scientific Publishing, Princeton.

NRC (National Research Council) (1983). *Risk assessment in the federal government: Managing the process*. National Academies Press, Washington, D.C.

NRC (National Research Council) (2009). *Science and decisions: Advancing risk assessment*. National Academies Press, Washington, D.C.

Regan, M. M. and Catalano, P. J. (1999). Likelihood models for binary and continuous outcomes: Application to Developmental Toxicology. *Biometrics*, 55, 760–768.

Slob, W. and Pieters, M. N. (1998). A probabilistic approach for deriving acceptable human intake limits and human health risks from toxicological studies: General framework. *Risk Analysis*, 18, 787–798.

Torres, J. and Bobst, S. (2015). *Toxicological risk assessment for beginners*. Springer, New York.

US DHHS (Department of Health and Human Services) (1985). Risk assessment and risk management of toxic substances. A report to the Secretary from the Executive Committee. DHHS Committee to Coordinate Environmental and Related Programs.

US EPA (Environmental Protection Agency) (1996). Report on the benchmark dose peer consultation workshop. Risk Assessment Forum, Washington, D.C., EPA/630/R-96/011.

US EPA (Environmental Protection Agency) (2005). Guidelines for carcinogen risk assessment. Risk Assessment Forum, Washington, D.C., EPA/630/P-03/001B.

US EPA (Environmental Protection Agency) (2011). Toxicological review of trichloroethylene, appendix G. In support of Summary Information on the Integrated Risk Information System (IRIS). Washington, D.C., EPA/635/R-09/011F.

US EPA (Environmental Protection Agency) (2012). Benchmark dose technical guidance. Risk Assessment Forum, Washington, D.C., EPA/100/R-12/001.

West, R. W. and Kodell, R. L. (1999). A comparison of methods of benchmark-dose estimation for continuous response data. *Risk Analysis*, 19, 453–459.

Yu, Zi-Fan and Catalano, P. J. (2005). Quantitative risk assessment for multivariate continuous outcomes with application to neurotoxicology: The bivariate case. *Biometrics*, 61, 757–766.

3

Statistical Models for Chemical Mixtures

3.1 Introduction

Although most of the toxicological research for assessing toxicity of chemicals and their effects on organisms and humans has focused on exposure to only a single chemical, the fact is that in reality humans are seldom exposed to one compound. More realistically, we concomitantly are exposed to a multitude of chemicals either concurrently or sequentially. Exposure to industrial and municipal waste, pesticides, herbicides, and gasoline are some examples of airborne unintentional exposure. Similarly, food additives and medicines that we consume contain a mixture of several components. Theoretically, there are an infinite number of combinations of two or more compounds and clearly testing for every combination is impossible. For these reasons, scientists have, for a long time, considered analyzing the combined effect of mixtures of chemicals based on the knowledge on the effects of the individual compounds comprising the mixture. In fact, the idea dates back to the late 1920s when Loewe and Muischnek (1926) discussed the notion of concentration addition or dose addition. Since then, a large body of publications, including several books and papers, has discussed various statistical models that characterize the joint actions of two or more chemicals. This chapter, therefore, is devoted to describing some of the more common models that can be utilized to predict the effects of chemical mixtures from the knowledge on the effects of single chemicals. For simplicity, attention is first focused on characterizing the joint action of two chemicals or the so-called binary mixtures. We then discuss the more general case of the combined effect of any finite number of chemicals.

For the purpose of mixture risk assessment, generally two approaches are possible. One approach is the so-called whole mixture approach where the mixture is treated as a single chemical. Similar to what we described in the last chapter, and just like risk assessment procedures for one chemical, first a dose-response relationship is established and is used to characterize the risk. The difficulty with this approach, as outlined by Lipscomb et al. (2010), is the uncertainty about the chemical composition of the mixture to which humans are exposed. The US Environmental Protection Agency (EPA, 2000) provides guidelines for conducting risk assessment for the "whole mixture" approach and addresses the uncertainties of this approach. More commonly, the second approach is the component-based approach and the methodology of risk assessment depends on whether or not the components can be classified as toxicologically similar, independent, or a combination of both. This will, in turn, determine how individual risks from each compound may be combined to give a risk estimate for the mixture. The complexity of characterizing the joint action of two or more drugs stems from the fact that toxicities of drug combinations involve not merely the enhancement of the potency of one drug with another, but also an interaction between the drugs. Interaction of two or more chemicals can produce one or more new chemical compound or can alter the properties of the mixed chemicals. If two chemical have no interaction, we call them *noninteractive*. The joint effect of noninteractive chemicals is additive and here we first discuss the additive models. We then focus on characterizing the interaction of chemicals and detection of departure from additivity.

3.2 Additivity Models for Noninteractive Chemicals

The interesting fact about mixtures is that often in the low-dose region the joint effect of two chemical agents can be assumed to additive. However, the nature of additivity depends on the characteristics of the compounds in the mixture. There are fundamentally two types of additivity models. One is *concentration additivity* or simply *dose additivity* and the other is *response additivity*. The terminology was introduced by Shelton and Weber (1981). Earlier, Loewe (1953) utilized the terms iso-addition and hetero-addition to describe the general ideas of additivity. Barenbaum (1985, 1989) argues that dose addition is the only correct basis for assessing the joint action of chemicals with no interaction and that response addition is simply a special case of dose additivity. The decision as to which type of additivity can be applied in a specific setting depends on the mode of action of the chemicals in the mixture and is based on either data on each component or expert opinion. If two chemicals have a similar mode of action and have essentially the same toxicological effects, then one chemical can be considered to be a dilution of the other one and dose addition is used. On the other hand if two chemicals react independently and have independent toxic effects, then response additivity may be applied. Let $P(d_i)$ be the probability of a toxic response to a dose d_i of toxicant i; $i = 1, 2$ and assume that $F_i(d)$; $i = 1, 2$ are the dose-response functions with F_i being some appropriate monotonic function such as the logistic or probit models. Then the probability of a toxic response to the combination of dose d_1 of chemical 1 and dose d_2 of chemical 2 under dose additivity can be expressed as

$$P(d_1, d_2) = F_1(d_1 + \rho\, d_2) = F_2\left(\frac{d_1}{\rho} + d_2\right) \tag{3.1}$$

where ρ is the potency of toxicant 2 relative to potency of toxicant 1 and is called the Relative Potency Factor. It is determined as the ratio of dose levels of the two chemicals producing the same levels of response. If, on the other hand, the effect of the two chemicals have no common mode of action and they act independently, then response additivity is applied and the probability of the joint action of the mixture of two chemicals with dose d_1 of chemical 1 and dose d_2 of chemical 2 is given by

$$P(d_1, d_2) = F_1(d_1) + F_2(d_2) - F_1(d_1).F_2(d_2) \tag{3.2}$$

based on the probability of the union of two independent events. In practice, the cross product term is negligible at low doses and is often ignored.

For the statistical analysis of the joint action of a mixture of two noninteractive compounds, Kodell and Pounds (1991) recommend that linearizing dose-response models $F_1(.)$ and $F_2(.)$ are selected so that the probability of an adverse effect from each chemical can be characterized as either a linear function of the dosage level or its logarithmic transformation, i.e. either

$$P(d_i) = F_i(\beta_{0i} + \beta_{1i} d_i) \quad i = 1, 2 \tag{3.3}$$

or

$$P(d_i) = F_i(\beta_{0i} + \beta_{1i} \log d_i) \quad i = 1, 2$$

where the model parameters are estimated using linear regression. Similarly, a linear regression is performed either on the dosage or the logarithm of the dosage for the mixture. Under concentration additivity, from (3.1) we have

$$P(d_1, d_2) = F_1\left\{\beta_{01} + \beta_{11}\left(d_1 + \rho d_2\right)\right\} = F_2\left\{\beta_{02} + \beta_{12}\left(\frac{d_1}{\rho} + d_2\right)\right\} \tag{3.4}$$

which holds true for every value of d_1 and d_2 and thus if F_1 and F_2 are from the same family of dose-response models, we have $\beta_{01} = \beta_{02} = \beta_0$ (say) and

$$d_1\left(\beta_{11} - \frac{\beta_{12}}{\rho}\right) + d_2\left(\rho\, \beta_{11} + \beta_{12}\right)$$

which gives $\rho = \dfrac{\beta_{12}}{\beta_{11}}$. Therefore, a combined regression may be performed on d_1 and d_2 using (3.5) to estimate the common intercept β_0, the relative potency factor ρ, and one of the slope parameters β_{11} or β_{12}. Using a combined regression on the logarithm of dose, on the other hand, the assumption of concentration addition yields

$$P(d_1, d_2) = F_1 \left\{ \beta_{01} + \beta_{11} \log \left(d_1 + \rho \, d_2 \right) \right\} = F_2 \left\{ \beta_{02} + \beta_{12} \log \left(\frac{d_1}{\rho} + d_2 \right) \right\} \quad (3.5)$$

and by setting the arguments equal, we get

$$\frac{\left(\dfrac{d_1}{\rho} + d_2 \right)^{\beta_{12}}}{\left(d_1 + \rho \, d_2 \right)^{\beta_{11}}} = e^{-(\beta_{02} - \beta_{01})} \quad (3.6)$$

which, by writing it as

$$\left(d_1 + \rho \, d_2 \right)^{(\beta_{12} - \beta_{11})} = \rho^{\beta_{12}} e^{-(\beta_{02} - \beta_{01})}$$

clearly leads to $\beta_{11} = \beta_{12} = \beta_1$ (say) and $\rho = e^{(\beta_{02} - \beta_{01})/\beta_1}$. Thus the joint response may be predicted by regressing the response on log-dose of the mixture and estimating the common slope β_1, the relative potency ρ, and one of the intercept parameters β_{01} or β_{02}.

Now, if the chemicals are independent, then under response additivity, we have

$$P(d_1, d_2) = F_1 \left(\beta_{01} + \beta_{11} d_1 \right) + F_2 \left(\beta_{02} + \beta_{12} d_2 \right)$$
$$- F_1 \left(\beta_{01} + \beta_{11} d_i \right) \cdot F_2 \left(\beta_{02} + \beta_{1i} d_2 \right) \quad (3.7)$$

for the dose and

$$P(d_1, d_2) = F_1 \left(\beta_{01} + \beta_{11} \log d_1 \right) + F_2 \left(\beta_{02} + \beta_{12} \log d_2 \right)$$
$$- F_1 \left(\beta_{01} + \beta_{11} \log d_1 \right) \cdot F_2 \left(\beta_{02} + \beta_{1i} \log d_2 \right) \quad (3.8)$$

for the log-dose. For the choice between using the actual dose or the logarithmic transformation of the dose, Kodell and Pounds (1991) suggest using a χ^2 type goodness-of-fit test if the responses are quantal and a F-test based on the ratio of the between and within dose group sums of squares of mixture responses and select the more appropriate model.

It is also worth noting that if the dose-response model is the exponential distribution, representing the one-hit model with zero background,

$$P(d_i) = 1 - \exp\left(-\lambda_i d_i \right) \quad i = 1, 2$$

then from (3.1) and (3.2) both concentration additivity and response additivity yield the same function

$$P(d_1, d_2) = 1 - \exp\left[-\left(\lambda_1 d_1 + 2 d_2\right)\right]$$

and become equivalent. Similarly, if the dose-response function is strictly linear, then for low doses the two types of additivity become equivalent.

Although the above approach for estimating the joint action of a mixture of two noninteractive chemicals is quite reasonable, the choice between the actual dose and the log-dose for using in the linearized transformation remains rather arbitrary. Razzaghi and Kodell (1992) argue that a more reliable modeling could be achieved by allowing the data to choose the right transformation of the data by the Box–Cox transformation that provides the best fit. This will in turn account for the fact that the evaluation of the dose-response models with log-transformed doses becomes indeterminate in the background. The logarithmic transformation is more naturally included in the larger and richer family of Box–Cox transformation. Suppose that instead of using the concentration d_i or the log concentration $\log d_i$ in (3.3) and (3.4), we define the probability of a toxic response at dose d_i of chemical i; $i = 1, 2$ is defined by

$$P(d_i) = F_i\left(\beta_{0i} + \beta_{1i} \frac{d_i^{\gamma_i} - 1}{\gamma_i}\right) \quad i = 1, 2 \tag{3.9}$$

where γ_1 and γ_2 are parameters to be estimated in such a way that the most suitable power transformations for the concentrations d_1 and d_2 are derived. The inclusion of the logarithmic transformation is accomplished by noticing that the Box–Cox family of transformation (3.7) is continuous in γ_i and when $\gamma_i \to 0$, the logarithm resulted. Thus small values of γ_i in the Box–Cox transformation (3.2) represent transformations that are close to logarithmic transformation. As γ_i increases, a whole family of transformation is obtained and the limiting case of $\gamma_i \to 1$ gives the actual dose. In other words using the concentration or the log concentration can be considered as the two extreme cases and by allowing γ_i to vary in [0,1] the most suitable power transformation can be selected. As is well known, the importance of the Box–Cox transformation is that it is an instance of a situation where the direction of data analysis is determined by the data.

Now, using (3.9), under the assumption of concentration additivity, we have

$$P(d_1, d_2) = F_1\left\{\beta_{01} + \beta_{11} \frac{\left(d_1 + \rho d_2\right)^{\gamma_1} - 1}{\gamma_1}\right\}$$

$$= F_2\left\{\beta_{02} + \beta_{12} \frac{\left(\dfrac{d_1}{\rho} + d_2\right)^{\gamma_2} - 1}{\gamma_2}\right\} \tag{3.10}$$

from which we get

$$\beta_{01} - \beta_{02} = \frac{\beta_{11}}{\gamma_1} - \frac{\beta_{12}}{\gamma_2}$$

and

$$\beta_{11} \frac{1 - \rho^{\gamma_1}}{\gamma_1} + \beta_{12} \frac{1 - \left(\frac{1}{\rho}\right)^{\gamma_2}}{\gamma_2} = 0$$

and after some manipulation, we have $\gamma_1 = \gamma_2 = \gamma$ (say), meaning that the same power transformation of the concentration provides the best modeling structure for both chemicals leading to

$$\rho = \sqrt{\left(\beta_{12}\big/\beta_{11}\right)} \tag{3.11}$$

and

$$\gamma = \frac{\beta_{12} - \beta_{11}}{\beta_{02} - \beta_{01}}. \tag{3.12}$$

Thus, the statistical analysis under the assumption of concentration additivity consists of a combined regression of the transformed concentration with constraints (3.11) and (3.12). Clearly, a nonlinear regression procedure must be used in order to estimate the parameters including the joint Box–Cox transformation parameter γ. Interestingly in (3.12), $\gamma \to 0$ results in a common slope and $\gamma \to 1$ results in equality of intercepts, demonstrating once again that the current approach includes the two previous models as special cases.

Under the assumption of response additivity, the Box–Cox modeling approach yields

$$P(d_1, d_2) = F_1\left(\beta_{01} + \beta_{11}\frac{d_1^{\gamma_1} - 1}{\gamma_1}\right) + F_2\left(\beta_{02} + \beta_{12}\frac{d_2^{\gamma_2} - 1}{\gamma_2}\right)$$

$$- F_1\left(\beta_{01} + \beta_{11}\frac{d_1^{\gamma_1} - 1}{\gamma_1}\right) F_2\left(\beta_{02} + \beta_{12}\frac{d_2^{\gamma_2} - 1}{\gamma_2}\right) \tag{3.13}$$

with no particular restriction on the parameters and thus, using nonlinear regression, all parameters are free and need to be estimated.

Example 3.1

Carter et al. (1988) describe an experiment designed to study the combined effect of a mixture of ethanol and chloral hydrate in female mice. A total of 234 female ICR mice were randomly divided into 39 experimental dose groups with each group containing 6 mice. A 5×5 full factorial design was used for the mixture at five concentration levels for each chemical making a combination of 25 levels. In addition, seven dosages of each chemical in isolation were used. The mice were treated by the appropriate levels of the chemical through injection and response of interest was loss of righting reflex at 30 min after treatment. Table 3.1 provides the information on the dosage level for each chemical as well as number of responses out of 6.

For the purpose of illustration of the methodology we use the probit link function, although other linearizing links such as the logistic may be used. The probit link is based on the cumulative distribution function of the standard normal distribution. Thus we assume that $F_1(.) = F_2(.) = \Phi(.)$, cumulative distribution function of the standard normal distribution. Using the *glm* function for generalized linear models in R to fit a probit regression on ethanol alone, we find that the estimates for the intercept and slope for chemical 1 (ethanol) alone are

$$\hat{\beta}_{01} = -13.4914 \text{ and } \hat{\beta}_{11} = 0.0394$$

for concentration and

$$\hat{\beta}_{01} = -79.983 \text{ and } \hat{\beta}_{11} = 13.714$$

for log concentration respectively. Both models provided adequate fit to the data judging by the very similar values (11.542 and 11.429) of the Akaike Information Criterion (AIC). In fact a Kolmogorov-Smirnov goodness-of-fit test did not reject the adequacy of either model with identical *p*-values ($p = 0.5412$). Similarly, using the probit link function for regression of responses on chloral hydrate, we find that the estimates of the intercept and slope for chemical 2 (chloral hydrate) are:

$$\hat{\beta}_{01} = -18.1378 \text{ and } \hat{\beta}_{11} = 0.00439$$

for concentration and

$$\hat{\beta}_{02} = -151.477 \text{ and } \hat{\beta}_{12} = 18.191$$

for log concentration respectively. Once again, both models produced very similar fits judging by the AIC values (20.658 and 20.660) and the Kolmogorov-Smirnov goodness-of-fit test did not reject the adequacy of either model with identical *p*-values ($p = 0.9375$).

Turning our attention to the mixture of the two chemicals now, under concentration addition assumption using (3.4) and the *glm* function in R, we find that the estimates of the common intercept and the two slopes are respectively

$$\hat{\beta}_0 = -0.6056, \ \hat{\beta}_{11} = 0.0006179, \text{ and } \hat{\beta}_{11} = 0.0001619$$

TABLE 3.1

Toxicity Effect of Mixture of Ethanol–Chloral Hydrate in Mice

Ethanol mg/kg	Chloral Hydrate mg/kg	Number of Responses/6
0	300	0
0	325	2
0	350	4
0	375	5
0	400	6
0	425	6
0	450	6
4000	0	2
4050	0	2
4100	0	2
4150	0	3
4200	0	4
4250	0	5
4300	0	4
200	100	0
200	150	0
200	200	0
200	250	0
200	300	3
900	100	0
900	150	0
900	200	0
900	250	6
900	300	4
1600	100	0
1600	150	0
1600	200	5
1600	250	5
1600	300	6
2300	100	0
2300	150	4
2300	200	6
2300	250	6
2300	300	6
3000	100	6
3000	150	6
3000	200	6
3000	250	6
3000	300	6

leading to an estimate of $\hat{\rho} = 0.0001619/0.0006179 = 0.2620$ for the relative potency factor. Now, to fit the log concentration under concentration additivity, from (3.5) it is clear that we do not have a linear function of the explanatory variable in the regression and therefore the *glm* function in R cannot be used. For our purpose, *bnlr* for binomial nonlinear binomial regression may be used. The function resides in the R package *gnlm*. In this case, we find that the estimates of the two intercepts and the common slope are respectively

$$\hat{\beta}_{01} = -3.8453, \quad \hat{\beta}_{02} = -8.4054, \quad \text{and} \quad \hat{\beta}_1 = 0.5947$$

resulting in an estimate of $\hat{\rho} = \exp\left(\dfrac{-2.4741 + 2.0574}{0.2452}\right) = 0.1606$ for the relative potency factor.

Under the assumption of response additivity, either of the equations (3.7) or (3.8) may be used with the associated parameter estimates.

This dichotomy and uncertainty about the models can be eliminated by using the Box–Cox transformation to select the most appropriate power transformation. Using (3.9) separately for each chemical, ethanol and chloral hydrate, and using the function *bnlr* in R to fit a nonlinear binomial regression with a probit link function, we find that the parameter estimates are

$$\hat{\beta}_{01} = -0.7509, \quad \hat{\beta}_{11} = 0.1844, \quad \hat{\gamma}_1 = 0.0209$$

for ethanol and

$$\hat{\beta}_{02} = -0.7374, \quad \hat{\beta}_{12} = 0.0895, \quad \hat{\gamma}_2 = 0.0085$$

giving appropriate power transformation for each regression. Now, for the mixture, under the assumption of concentration additivity, using (3.10) with a probit link, the parameter estimates are

$$\hat{\beta}_{01} = -3.8453, \quad \hat{\beta}_{11} = 0.5945, \quad \hat{\gamma} = 0.0209, \quad \hat{\rho} = 0.0189$$

and from restrictions (3.11) and (3.12) the parameters $\hat{\beta}_{02}$ and $\hat{\beta}_{12}$ may be derived. The risk estimates for the joint action at any dose level can be estimated using (3.10). Under the assumption of response additivity, (3.13) is used to find risk estimates for the joint action.

3.3 Mixture of Interactive Chemicals

At the outset, we are most likely to be exposed to a mixture of several compounds. The substances polluting the air or water, food additives, and medicinal compounds generally contain more than two components. What makes the analysis of chemical mixtures complicated is the fact that chemicals often interact and their joint effect is not simply additive. In order to better understand the statistical modeling and risk assessment methods for complex mixtures, we first discuss and illustrate these concepts for mixtures that consist of only two chemicals. The first question that one needs to consider when mixing two chemicals is how the two compounds react in combination. Thus the most fundamental question is the characterization of the interaction of the two chemicals.

3.3.1 Characterizing Interactions of Two Chemicals (Binary Mixtures)

Determining how the components of a mixture interact and characterizing the nature of the interaction is an important step in toxicity assessment of a mixture. In general, we say two drugs are synergistic if the effect of the mixture of the two chemicals is greater than the sum of the individual effects at the same dose levels. Conversely, we say two drugs interact antagonistically if the effect of the combination of the two drugs is lower than the sum of the effects of the individual drugs at the same exposure levels. If two drugs are neither synergistic nor antagonistic, they are noninteractive and thus their effects are additive. Identifying how two chemicals react in combination is an old problem. The traditional and classical approach to detect interaction is a graphical method and relies on the use of an isobologram. An isobologram, introduced first by Fraser (1872), is a plot of a contour of the mixture dose-response surface for a fixed-effect level superimposed on the contour of the same level of response with the assumption of additivity. The fixed level of effect is selected as a dosage combination that affects $100\, p_0\%$ of the subjects with $0 \le p_0 \le 1$. Often p_0 is taken to be 0.5. Thus the line of additivity is the line that connects the ED_{100p_0} for chemical 1 to that of chemical 2. If the graph of the joint action of the chemicals is at least approximately on the line of additivity, the two chemicals are noninteractive. If, on the other hand, the graph of the concentration levels of the two chemicals for a fixed response is sufficiently above the additivity line, that indicates antagonism and if the graph falls below the line of additivity, it shows that the two chemicals are synergistic. Figure 3.1 is an illustration of the isobologram for two chemicals with ED_{50} as the fixed level of effect. Clearly, the application of this graphical approach has many limitations. Not only is its use limited to a mixture of two or at most three chemicals, most importantly, it does not account for the variability in responses. Moreover, the methodology does not reward higher sample sizes and the decision for characterizing the type of interaction between additivity, synergism, or antagonism does not incorporate a level of significance associated with the decision. In the absence of any formal statistical methodology for detecting departure from linearity and existence of synergism or antagonism, Gessner and Cabana (1970) suggest constructing pointwise confidence intervals at each combination of drug concentrations

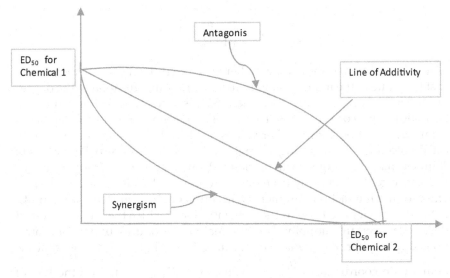

FIGURE 3.1
Illustration of an isobologram.

in the experiment and if the line of additivity crosses one or more of the confidence bands, then the chemicals can be assumed to be nonadditive. If, on the other hand, the line of additivity crosses at least one of the confidence bands then either synergism or antagonism can be claimed depending on the shape of the contour as described before. Because of the multiple testing, this method, however, suffers from an inflated type 1 error and a false-positive conclusion. To protect against this inflated type 1 error, often an α sharing procedure such as the Bonferroni correction or a less conservative approach such as the Holm's adjustment method is used (see for example Miller, 1991). However, such adjustment should be used with caution as the number of tests could be far beyond the recommended values in practice.

Note that analytically, the equation of the line of additivity for a dose d_1 of chemical 1 and dose d_2 of chemical 2 can be expressed as

$$\frac{d_1}{\mathrm{ED}^{(1)}_{100p_0}} + \frac{d_2}{\mathrm{ED}^{(2)}_{100p_0}} = 1 \tag{3.14}$$

where $\mathrm{ED}^{(1)}_{100p_0}$ and $\mathrm{ED}^{(2)}_{100p_0}$ are respectively the ED_{100p_0} for chemicals 1 and 2. Thus in practice if the equality in (3.14) holds, the two chemicals may be declared as additive, whereas if the sum on the left-hand side is more than unity, then supra-linearity is indicated and the two chemical agents are antagonistic. Similarly, if the sum on the left side of (3.14) is less than 1, then the graph is sub-linear, indicating synergism. Hewlett (1969) introduced a generalization of (3.14) by adding an interaction parameter λ. That expression is given by

$$\left(\frac{d_1}{\mathrm{ED}_{100po}^{(1)}}\right)^{1/\lambda} + \left(\frac{d_2}{\mathrm{ED}_{100po}^{(2)}}\right)^{1/\lambda} = 1 \tag{3.15}$$

which, in addition to showing whether or not two chemicals interact, shows that the value of the interaction parameter λ gives an indication of the degree of interactivity. If λ is equal to 1, then (3.14) is resulted showing concentration additivity. For other positive values, λ gives the degree of departure from concentration addition. For values of λ less than 1, antagonism is indicated with interaction getting stronger as λ gets closer to zero. For values of λ higher than 1, synergism can be claimed, once again with stronger interaction as we get further away from one. Sorensen et al. (2007) give an interpretation of the interaction parameter λ in terms of the sum of the toxic units, defined in binary mixtures as the concentration at which there is a 50% effect (EC50) for a certain endpoint. For 1:1 equal ratio mixtures, under concentration additivity for noninteractive mixtures, from Figure 3.2, we see that the points with coordinates $\left(\dfrac{\theta_1}{2}, \dfrac{\theta_2}{2}\right)$ where $\theta_i = \mathrm{ED}_{100po}^{(i)}$; $i = 1, 2$ on the line of additivity produces the same effect as the points with coordinates $(\theta_1, 0)$ and $(0, \theta_2)$, each leading to a value of $\lambda = 1$. If, on the other hand, the chemicals are interactive and not additive, say if they are synergistic, then we see from Figure 3.2 that the point with coordinates $\left(\theta_1^*, \theta_2^*\right)$ produces the same effect. Thus from (3.15), we have

$$\left(\frac{\theta_1^*}{\theta_1}\right)^{1/\lambda} + \left(\frac{\theta_2^*}{\theta_2}\right)^{1/\lambda} = 1 \tag{3.16}$$

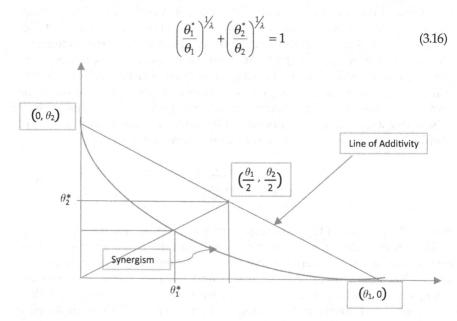

FIGURE 3.2
Interpreting the interaction parameter.

which, by symmetry, indicates that

$$\left(\frac{\theta_i^*}{\theta_i}\right)^{1/\lambda} = \frac{1}{2} \quad i = 1, 2$$

and thus,

$$\theta_i^* = \frac{\theta_i}{2^\lambda} \quad i = 1, 2.$$

Therefore, the sum of the toxic units is given by

$$\frac{\theta_1^*}{\theta_1} + \frac{\theta_2^*}{\theta_2} = 2^{1-\lambda} \tag{3.17}$$

This means that 50% effect of the biological endpoint of interest is achieved by a dose of the mixture which is $2^{1-\lambda}$ times the dose of each chemical to achieve the same effect under concentration additivity. For example, if the two chemicals are synergistic and λ is estimated as 1.7, say, then $2^{1-\lambda} \approx 0.62$. Similarly, if the two chemicals are antagonistic and $\lambda = 0.5$, then $2^{1-\lambda} = 1.4$. As noted in Sorensen et al. (2007), although (3.17) is for 1:1 equal ratio mixtures, it is possible to derive similar expressions for mixtures with other ratios as well, but the corresponding expressions become more complicated. For example, if the two chemicals are in the ratio of 1:3, then from (3.16)

$$\left(\frac{\theta_1^*}{\theta_1}\right)^{1/\lambda} = \frac{1}{4}$$

and

$$\left(\frac{\theta_2^*}{\theta_2}\right)^{1/\lambda} = \frac{3}{4}$$

leading to

$$\frac{\theta_1^*}{\theta_1} + \frac{\theta_2^*}{\theta_2} = 2^{-\lambda}(1 + 3^\lambda)$$

as the sum of the toxic units.

Noting that (3.15) assumes symmetry in the isobole, Hewlett (1969) introduced a more flexible model with two interaction parameters as

$$\left(\frac{d_1}{ED_{100p_0}^{(1)}}\right)^{1/\lambda_1} + \left(\frac{d_2}{ED_{100p_0}^{(2)}}\right)^{1/\lambda_2} = 1.$$

However, this model does not have a simple interpretation of the interaction parameters λ_1 and λ_3 similar to (3.15). Other generalizations are also given by Volund (1992) and Streibig et al. (1998).

3.3.2 Statistical Methods for Departure from Linearity of Two Chemicals

To incorporate statistical methods in the analysis of the isobologram, Carter et al. (1988) introduce an approach based on response surface methodology. Let $P(d_i)$ be the probability of a toxic response to a dose d_i of toxicant i; $i = 1, 2$. Similarly, let $P(d_1, d_2)$ be the probability of a toxic response to the combination of dose d_1 of chemical 1 and dose d_2 of chemical 2. Assume that $F_1(.)$, $F_2(.)$, and $F(.)$ are the corresponding appropriate monotonic dose-response functions such as the logistic or probit models for the two compounds and the mixture respectively. Then, using a generalized linear model, we can write

$$g^{-1}\left(\beta_0 + \beta_1 d_1 + \beta_2 d_2 + \beta_{12} d_1 d_2\right) = p_0 \tag{3.18}$$

where β_0, β_2, and β_{12} are unknown parameters and $g(.) = F^{-1}(.)$ serves as the link function (see McCullagh and Nelder, 1989). For example, if the dose-response function of the mixture can be expressed by the logistic model, we have

$$\operatorname{logit}(p_0) = \log\left(\frac{p_0}{1 - p_0}\right) = \beta_0 + \beta_1 d_1 + \beta_2 d_2 + \beta_{12} d_1 d_2. \tag{3.19}$$

From (3.18), in the absence of chemical 2, by substituting $d_2 = 0$, for chemical 1 alone we specifically have

$$\beta_0 + \beta_1 \mathrm{ED}^{(1)}_{100\alpha} = g(p_0)$$

or

$$\mathrm{ED}^{(1)}_{100\alpha} = \frac{g(p_0) - \beta_0}{\beta_1}. \tag{3.20}$$

Similarly, by substituting $d_1 = 0$ in (3.18), for chemical 2 alone we have

$$\mathrm{ED}^{(2)}_{100\alpha} = \frac{g(p_0) - \beta_0}{\beta_2}. \tag{3.21}$$

Now, writing (3.18) as

$$\frac{\beta_1 d_1}{g(p_0) - \beta_0} + \frac{\beta_2 d_2}{g(p_0) - \beta_0} + \frac{\beta_{12} d_1 d_2}{g(p_0) - \beta_0} = 1$$

we have

$$\frac{d_1}{\left(g(p_0)-\beta_0\right)/\beta_1} + \frac{d_2}{\left(g(p_0)-\beta_0\right)/\beta_2} = 1 - \frac{\beta_{12}d_1d_2}{\left(g(p_0)-\beta_0\right)}$$

and by using (3.20) and (3.21) we get

$$\frac{d_1}{\mathrm{ED}_{100p_0}^{(1)}} + \frac{d_2}{\mathrm{ED}_{100p_0}^{(2)}} = 1 - \frac{\beta_{12}d_1d_2}{\left(F^{-1}(p_0)-\beta_0\right)}. \tag{3.22}$$

Since $\left(g(p_0)-\beta_0\right) > 0$ as long as the level of response p_0 is higher than the background response level $F(\beta_0)$, then by comparing (3.14) and (3.22) we can deduce that the two chemicals interact synergistically if $\beta_{12} > 0$. Similarly, for $\beta_{12} < 0$, the chemicals interact antagonistically and they are additive when the coefficient of the cross product term is zero, i.e. when $\beta_{12} = 0$. This analytical interpretation of the isobologram has interesting and important implications. Not only does it allow for maximum likelihood estimation of the model parameters and goodness-of-fit testing procedures, most importantly, the hypothesis of linearity can directly be tested based on the parameter β_{12}. Note that it is also possible for two chemicals to be synergistic in a certain dose region, becoming antagonistic in higher doses or vice versa. In those situations, it is possible to extend the methodology described above to include more complex functions of the linear predictor in (3.18) that includes second-order terms or higher. We will not discuss those methods here and refer the reader to Carter et al. (1988).

Example 3.2

In Example 3.1, we described an experiment to study the toxic effect of a mixture of ethanol and chloral hydrate. Assuming no interaction, we derived dose-response models for the joint action of the two chemicals. Carter et al. (1988) applied the methodology described in Section 3.1 above along with a logistic link function to show that the two chemicals are not additive and have a synergistic effect. Here, we use the probit link to investigate nonadditivity and the type of interaction of the two drugs. Thus from (3.18) we let

$$\Phi^{-1}(\alpha) = \beta_0 + \beta_1 d_1 + \beta_2 d_2 + \beta_{12}d_1d_2.$$

Table 3.2 gives the parameter estimates along with their standard error and the p-values derived by using the probit link in the *glm* function in R for the generalized linear models. Note that the estimate of the coefficient of the cross product term β_{12} is positive and highly significant. This confirms that ethanol and chloral hydrate are not additive and indeed have a synergistic effect.

TABLE 3.2

Parameter Estimates for Interaction Model of a Mixture of Ethanol and Chloral Hydrate

Parameter	Estimate	Standard error	Significance
β_0	−0.4801	−1.648	0.00357
β_1	-1.422×10^{-4}	2.150×10^{-4}	0.508367
β_2	1.252×10^{-4}	6.221×10^{-5}	0.044216
β_{12}	3.986×10^{-6}	1.077×10^{-6}	0.000215

3.3.3 Mixtures of More Than Two Chemicals

The procedure described in the last section can easily be generalized to mixtures of any number of chemicals. Suppose that a mixture consists of k chemicals. Let $d_1, \ldots, d_k$ be a dose combination of the drugs that produces a mean response of $100p_0\%$, with $0 \leq p_0 \leq 1$, denoted by $\mathrm{ED}^{(j)}_{100p_0}$ $j = 1, \ldots, k$, the dosage of the jth component of the mixture that yields a mean response of $100p_0\%$ in isolation. Then, analogous to (3.14), the chemicals are additive (Gennings, 1995) when

$$\sum_{j=1}^{k} \frac{d_j}{\mathrm{ED}^{(j)}_{100p_0}} = 1 \qquad (3.23)$$

In the event that a chemical in isolation does not produce the required level of response $100p_0\%$, but has an effect in combination with other chemicals in the mixture, then the corresponding dosage in the denominator of (3.23) may be set to infinity so that the contribution of that chemical in (3.23) is zero. The left-hand side of (3.23) is called the additivity index and we declare the components of a mixture as noninteractive if the additivity index is equal to 1. If, on the other hand, the additivity index is less than 1, then synergism can be claimed. Similarly, if the additivity index is greater than 1, then antagonism can be claimed at the given dose combination. Suppose that $F(.)$ is the common known monotone linearizing transformation representing the dose-response function for the jth component in the mixture. That is, let the response y_j of the jth chemical in the mixture at dose d_j be characterized as

$$g^{-1}(y_j) = \beta_{0j} + \beta_{1j}d_j. \qquad (3.24)$$

Then, at the mean response level of $100p_0\%$, from (3.23) and (3.24) we have

$$\sum_{j=1}^{k} \frac{\beta_{1j}d_j}{F(p_0) - \beta_{0j}} = 1. \qquad (3.25)$$

We further assume that in the absence of any treatment, the control response for all the chemicals and the mixture is the same and denote the common background response by β_0. This assumption is clearly reasonable and leads to

$$\beta_{01} = \beta_{02} = \cdots = \beta_{0k} = \beta_0.$$

Then, (3.24) gives

$$g^{-1}(p_0) = \beta_0 + \sum_{j=1}^{k} \beta_j \, d_j \qquad (3.26)$$

which defines the additivity model. Using this additivity model, Gennings (1995) and Gennings and Carter (1995) describe a three-stage algorithm based on comparison of the estimated response under additivity assumption with the observed response of the mixture to detect departure from additivity at a given specific dose combination. Accordingly, suppose that there are C combinations of doses of the k components with $r_c; c = 1, \ldots, C$ measurements at dose combination c of the mixture and let $\bar{Y}_c; c = 1, \ldots,$ C be the mean response for dose combination c. Suppose m_i is the number of dose levels of the ith chemical in isolation in the experiment for $i = 1, \ldots, k$ and assume that $n_{ij}; j = 1, \ldots, m_i$ is the number of observations available at the jth dose level of the ith chemical. Let $Y_{ijl}, l = 1, \ldots, n_{ij}$ be the response of the lth subject in the jth dose group of the ith chemical. Then using the additivity model (3.26) since $F(.)$ is a known function, in the first stage of the algorithm model parameters are estimated. Gennings (1995) suggests using the square root function for $F(.)$ and method of iteratively reweighted least squares to estimate the parameters. But, clearly, other standard models such as logistic or probit functions can be used. Once the parameters in (3.26) are estimated, in the second stage a predicted value $\hat{Y}_c$ at a given dose combination c of the mixture can be determined and a prediction interval

$$\hat{Y}_c \pm t_{1-\frac{\alpha}{2}, \sum_{c=1}^{C} r_c - k} \sqrt{\left(\text{Var}\left(\bar{Y}_c\right) + \text{Var}\left(\hat{Y}_c\right) \right)} \qquad (3.27)$$

is constructed. In order to do so, an estimate of the prediction variance $\text{Var}\left(\bar{Y}_c\right) + \text{Var}\left(\hat{Y}_c\right)$ is required. Gennings and Carter (1995) describe a jack-knife resampling method due to Wu (1986) to determine the estimate of $\text{Var}\left(\hat{Y}_c\right)$. Now in stage 3, $\bar{Y}_c$ is used to decide whether or not the chemicals in the mixture can be claimed to be additive. If the value of $\bar{Y}_c$ falls in the interval (3.24), the chemicals are additive. However, if the value of $\bar{Y}_c$ is less than the lower bound of the interval, then the chemical at the dose combination c is antagonistic. Conversely, if the value of $\bar{Y}_c$ is higher than the upper bound in the interval (3.27), then the chemicals are synergistic. This methodology was further extended in Gennings et al. (1997) to include a threshold in the additivity model. In non-cancer risk assessment, it is commonly accepted that there is a threshold below which the response is the same as the background. This threshold is a function of the target tissue, the response being measured, and the statistical design of the experiment.

It should be emphasized that the conclusions are only local in that if it is concluded that the chemicals are additive at a given dose combination, they may not be so at other dose combinations. The prediction intervals are piecewise without correction for multiple testing and consequently will suffer from an inflated significance level for the family of tests. To correct this shortcoming, Gennings et al. (2002) propose an overall test of additivity based on a Wald-type test. Let $\bar{Y} = \left(\bar{Y}_1, \ldots, \bar{Y}_C\right)^T$ and $\hat{Y} = \left(\hat{Y}_1, \ldots, \hat{Y}_C\right)^T$ be respectively the observed mean vector and predicted vector under additivity for all the mixture points. Let Σ be the covariance matrix of the model parameters and let $T = \mathrm{diag}\left(\mathrm{Var}\left(\bar{Y}_1\right), \ldots, \mathrm{Var}\left(\bar{Y}_C\right)\right)$ be the $C \times C$ diagonal matrix with the ith diagonal element being the variance of $\bar{Y}_i$; $i = 1, \ldots, C$. Then by applying the multivariate version of the delta method, we can express the large sample variance–covariance matrix of $D = (\hat{Y} - \bar{Y})$ approximately as WSW^T where W is the matrix of partial derivatives of the elements of D with respect to each of the model parameters and

$$S = \begin{pmatrix} \Sigma & 0 \\ 0 & T \end{pmatrix}.$$

Then, for large samples, the statistic

$$U(D) = \frac{D^T \left(WSW^T\right)^{-1} D}{MSE} \tag{3.28}$$

follows an F distribution with C and $N - p$ degrees of freedom where N is the total number of observations and p is the total number of model parameters. Using (3.28) we can test the overall additivity hypothesis simultaneously. If the assumption of overall additivity is rejected, then it would be of interest to assess additivity or lack of it thereof at each of C mixture points. Gennings et al. (2002) suggest using a correction based on Hochberg's (1990) step-down sequentially rejective approach for single-degree-of-freedom tests. According to that approach, the significance level is corrected using a Bonferroni correction. The hypothesis of additivity at the cth mixture point is rejected when the p-value corresponding to the test is less than $\alpha / (C - c + 1)$; $c = 1, \ldots, C$. Alternatively, to construct the individual prediction interval at each mixture point, this correction is used to determine the corresponding significance level.

Example 3.3

Nesnow et al. (1998) describe an experiment designed and conducted in order to study the tumorigenic effect of a mixture of five environmental hydrocarbons. The chemicals under study were benzo[a]pyrene, benzo[b]fluoranthene, dibenz[a,h]anthracene, 5-methylchrysene, and cyclopental[c,d]pyrene, which are all the so called polycyclic aromatic hydrocarbons (PAHs). We refer to these chemicals in this example as chemicals 1–5. For more information about each chemical, we refer the reader to Nesnow et al. (1998). The male strain A/J mice 6–8 weeks of age were treated with doses of each chemical and a 2^5 full factorial experiment was used to assess the effect of the mixture. In such experiments, two levels (doses) of each factor (chemicals) commonly referred to as "low" and "high" levels are used in the experiment. Therefore there were a total of 32 combinations of doses for the mixture study. The outcome of interest was the number of tumors on the lungs of the surviving mice after 8 months. The experiment referred to as the "5 PAH mixture study" was conducted at the Medical College of Ohio and Ohio State University and was supported by the US EPA. Table 3.3 gives the concentration level along with the observed mean and variance of the number of tumors for each chemical and Table 3.4 gives the similar results for the 2^5 mixture experiment. Both tables are adapted from Gennings (1995) where the author uses a square root linearizing transformation, i.e. $F(p) = \sqrt{p}$ to examine departure from linearity. In this example we use the same data to demonstrate the methodology described in this section, but use the more traditional count model which is Poisson regression. For the purpose of illustration, we generate a representative site-specific data set using the given experimental means and variances. The details for generating the representative data set are given in Appendix 1.

Using the *glm* function in R, with a Poisson link, it was found that every chemical had a significant contribution to the response. Table 3.5 gives the estimate for the coefficients in the model for each chemical along with their associated standard deviations. A Wald-type test based on (3.28) strongly rejected the overall additivity. Using the parameter estimates, assumption of additivity can be examined at any combination of doses. Suppose, for example, that it is desired to examine additivity at dose combinations of 30, 75, 10, 30, 100 respectively for chemicals 1 through 5. Note from Table 3.4 that for this combination of drugs, the mean response is $\bar{Y}_c = 44.82$ with a variance of $\text{Var}(\bar{Y}_c) = 341.0$ from a sample of size 17. Now, using Table 3.5, we find that the expected number of responses is for this combination of doses is about $\hat{Y}_c = 203$ with an approximate standard deviation of 1.01. Therefore, a 95% prediction interval from (3.27) may be formed as $203 \pm t_{0.975,27} \sqrt{342}$ leading to the interval (165.09, 240.91). Since the observed response falls outside this predicted interval, the hypothesis of additivity is rejected and since the observed mean number of responses is less than the lower limit of the interval, it is reasonable to conclude that at the given dose combination, there

TABLE 3.3

Mean and Variance for Number of Lung Tumors in Mice

	Dose	Mean	Variance	Sample Size
Control	0	0.6	0.358	20
Chemical 1	5	0.45	0.682	20
	10	0.53	0.64	17
	50	4.37	7.91	19
	75	7.14	14.8	7
	100	12.7	19.5	16
	200	33	109.2	24
Chemical 2	10	0.67	0.588	18
	50	2	3.47	20
	75	4.1	1.43	10
	100	5.3	10.8	20
	200	6.95	13.1	19
Chemical 3	1.25	1.44	2.26	18
	2.5	3.05	3.83	19
	5	13.1	37.7	20
	10	28.7	123.5	30
Chemical 4	10	1.75	2.62	20
	30	13.5	54.5	10
	50	39	187.4	18
	100	93.1	398.6	15
Chemical 5	10	0.55	0.682	20
	50	4.75	4.72	20
	100	29.25	215.7	28
	200	97.7	868.1	19

Data from Gennings (1995).

is an overall antagonistic relationship between the five chemicals. But, as mentioned before, at any other dose combination this relationship may be different.

3.3.4 Application of 2^k Factorial Experiments

To detect interaction between two or more chemicals, a 2^k factorial design may be used. These experiments are very efficient and provide useful information about the interaction of factors at a low cost. As mentioned in Example 3.3, two levels of each factor commonly called "low" and "high" and denoted as "−" and "+" are used and a full factorial experiment consists of at least one trial for each possible combination of factors at both levels. Thus an experiment with two factors has four combinations, an experiment with three factors has eight combinations and so on. This kind of design may be used for mixtures of chemicals to detect any possible interactions between the chemicals in the mixture. For example, if a mixture consists of

TABLE 3.4

Mean and Variance of the Number of Tumors in the 2^5 Factorial Design Mixture Study

Chemical 1	Chemical 2	Chemical 3	Chemical 4	Chemical 5	Mean	Variance	Sample Size
30	30	2.5	10	30	18.66	158.1	15
30	30	2.5	10	100	31.75	241.5	16
30	30	2.5	30	30	29.50	67.9	20
30	30	2.5	30	100	47.06	208.8	15
30	30	10	10	30	26.22	107.1	18
30	30	10	10	100	42.41	257.5	17
30	30	10	30	30	40.86	532.1	15
30	30	10	30	100	57.78	362.8	14
30	75	2.5	10	30	16.80	82.0	20
30	75	2.5	10	100	37.68	298.0	16
30	75	2.5	30	30	28.90	120.8	20
30	75	2.5	30	100	39.21	176.1	19
30	75	10	10	30	25.95	78.8	20
30	75	10	10	100	32.27	75.0	18
30	75	10	30	30	38.95	256.4	19
30	75	10	30	100	44.82	341.0	17
75	30	2.5	10	30	26.71	162.6	17
75	30	2.5	10	100	30.12	174.8	16
75	30	2.5	30	30	34.21	206.1	19
75	30	2.5	30	100	45.35	226.4	17
75	30	10	10	30	36.37	199.0	16
75	30	10	10	100	45.29	282.3	17
75	30	10	30	30	50.44	451.6	18

TABLE 3.4 (CONTINUED)

Mean and Variance of the Number of Tumors in the 25 Factorial Design Mixture Study

Chemical 1	Chemical 2	Chemical 3	Chemical 4	Chemical 5	Mean	Variance	Sample Size
75	30	10	30	100	48.94	480.7	17
75	75	2.5	10	30	20.55	50.4	18
75	75	2.5	10	100	31.57	95.8	14
75	75	2.5	30	30	29.50	100.8	20
75	75	2.5	30	100	44.81	293.5	16
75	75	10	10	30	29.44	114.0	16
75	75	10	10	100	36.55	101.8	18
75	75	10	30	30	45.71	110.1	14
75	75	10	30	100	63.78	392.7	14

Data from Gennings (1995).

TABLE 3.5

Parameter Estimate for Poisson Regression

	Estimate	Standard Deviation	p-Value
Intercept	2.153	0.001662	< 0.0001
Chemical 1	0.006625	0.000011	< 0.0001
Chemical 2	−0.002349	0.000039	< 0.0001
Chemical 3	0.1201	0.000206	< 0.0001
Chemical 4	0.02423	0.00002	< 0.0001
Chemical 5	0.01211	0.000001	< 0.0001

TABLE 3.6

Design of a Three-Factor Two-Level Experiment

Treatment			
(1)	−	−	−
a	+	−	−
b	−	+	−
ab	+	+	−
c	−	−	+
ac	+	−	+
bc	−	+	+
Abc	+	+	+

three chemicals A, B, and C, the design structure is as displayed in Table 3.6 and the eight treatment combinations are conventionally denoted by (1), *a*, *b*, *ab*, *c*, *ac*, *bc*, *abc* where a lower case letter denotes the high level and its absence denotes the low level of the chemical. For example, *ac* refers to the treatment combination of low level of chemical B and high levels of chemicals A and C. The treatment combination (1) refers to low levels of all three chemicals. The advantage of such models is that not only can first-level interactions be tested, but higher-level interactions can be examined as well. Using linear regression and analysis of variance, the significance of interaction terms may be tested. The analysis proceeds by fitting a linear regression model and testing the significance of the coefficients. For example, for the case of a mixture with three chemicals, the generalized linear regression model

$$Y = \beta_0 + \beta_1 D_1 + \beta_2 D_2 + \beta_3 D_3 + \beta_{12} D_1 \times D_2$$
$$+ \beta_{13} D_1 \times D_3 + \beta_{23} D_2 \times D_3 + \beta_{123} D_1 \times D_2 \times D_3$$

where Y is the response and D_i $i = 1, 2, 3$ may without loss of generality be coded as −1 or +1 depending on whether chemical i is at a low level or high level is first fitted to the data. First- and second-level interactions are then tested using the corresponding test of significance on model parameters.

The 2^k factorial design becomes somewhat prohibitive and costly if the number of chemicals under study is rather large. For more information about factorial experiments, we refer to Montgomery (2017).

Example 3.4

The 5 PAH mixture study described in Example 3.3 was analyzed by Gennings (1995) using a 2^5 factorial design and by Nesnow et al. (1998) using a response surface methodology. Note that the model consists of the main effects as well as the first-, second-, third-, and fourth-level interactions. In the absence of site-specific data for the mixture effect, a representative data set may be generated using the method described in Example 3.2. We will not repeat the calculation here and refer the reader to Exercise 3 at the end of this chapter.

3.3.5 Fixed-Ratio Ray Mixtures

Factorial design is a very efficient and popular method to assess the effect of each factor individually and their interaction in combination with other chemicals. However, as discussed, the experiment becomes very costly, especially in mixture studies when the number of chemicals under study becomes rather large. For example, as illustrated in Example 3.3, in the 2^5 factorial experiment, 32 observations are required for each replication and if the experiment is replicated a few times, the experiment can become very large. Factorial experiments become even more prohibitive if a 3^k design is used. In such designs, three levels of each factor commonly referred to as "low", medium", and "high" are used. Thus in mixture studies, three doses of each chemical are crossed with three doses of all the other components of the mixture. Thus for example, each replication of a 3^5 experiment would require 243 runs. So, in effect, the application of factorial designs is limited to mixtures with a small number of chemicals. One way to resolve this problem is to use a fractional factorial experiment where only a subset of the treatment combinations is used. Although these experiments have the advantage of reducing the number of runs and thus making it possible for a larger number of chemicals to be examined, the disadvantage is that the effect of some chemicals and the interaction effect of other chemicals will become indistinguishable or 'aliased.' For more on this topic we refer to Montgomery (2017). In mixture studies an efficient method is to use a *fixed-ratio ray* design. Although the application of these designs goes back to the early 1940s, Meadows et al. (2002) provided a formal statistical analysis. Development of fixed-ratio ray designs alleviated many of the problems associated with the large number of trials in factorial experiments. A ray design is based on a relevant, fixed mixing ratio of interest $a_1, a_2, \ldots, a_k$ of the k chemicals where $\sum_{i=1}^{k} a_i = 1$ and allowing the total dose to vary. In this way, attention can be focused on inference along a particular fixed-ratio ray of interest. To test the hypothesis of additivity in fixed-ratio ray mixtures, we can use the same procedure described in Section 3.2 and determine if the observed mean responses from the mixture points coincide with the predicted responses under the additivity assumption by constructing a prediction interval at each experimental mixture point at which examination of additivity is desired. The difference, however, is that in Section 3.2 additivity could be examined at each combination of the concentration of the drug or chemical components in the factorial design mixture, whereas here we examine departure from additivity along a fixed-ratio ray of the components of the mixture.

Now, if we denote the total mixture dose by d, then the fraction of the total dose represented by the ith chemical component of the mixture is a_id. The design is efficient and useful for detecting departure from additivity even with a large number of chemicals. Methods for detecting departure involve modeling the response as a function of the total dose along the ray. For the response Y, and the link function $g(.)$, the additivity model may be expressed as

$$g(Y) = \beta_0 + \beta_1 d_1 + \cdots + \beta_k d_k + \epsilon \qquad (3.29)$$

where ϵ is the random error and d_i is the dosage of the ith component in the mixture which in the ray design is a_id. Substituting in (3.29), we have

$$g(Y) = \beta_0 + \left(\beta_1 a_1 + \beta_2 a_2 + \cdots + \beta_k a_k\right)d + \epsilon. \qquad (3.30)$$

Therefore in contrast to the factorial design, in the fixed-ratio ray design, the dose-response relationship under additivity can be expressed as a single generalized linear model on a total dose with a slope of

$$\beta^* = \sum_{i=1}^{k} a_i \beta_i. \qquad (3.31)$$

This feature makes the overall test of additivity much more amenable for fixed-ratio ray mixtures.

3.3.6 Test of Additivity for Fixed-Ratio Ray Mixtures

From (3.31), it is apparent that if the slope of a simple regression on total dose is equal to $\sum_{i=1}^{k} a_i \beta_i$, then the additivity equation given by (3.23) holds. Hence, to test for additivity, single dose-response models are first fitted to the data for each individual chemical and a single dose-response data is used on the total dose. If (3.31) holds, then the hypothesis of linearity is not rejected. Following Casey et al. (2004), we can formulate a test of hypothesis for additivity as follows. Let L be the total number of fixed-ratio rays of interest and let $a_{(l)} = \left(a_{1(l)}, a_{2(l)}, \ldots, a_{k(l)}\right)^T$ be the mixing ratios along the *l*th fixed-ratio ray for $l = 1, \ldots, L$. Suppose that $d_{(l)} = \left(d_{1(l)}, d_{2(l)}, \ldots, d_{k(l)}\right)^T$ is the vector of the dose levels of the k components of the mixture and let $d = \sum_{i=1}^{k} d_{i(l)}$ be the total dose along the *l*th ray. Thus the amount of the *i*th chemical in the mixture along the *l*th ray is $a_{i(l)}d$; $i = 1, \ldots, k$. We now use $g(.)$ as the link function and fit the linear model (3.29) to each individual chemical along each ray. Similarly, a linear model

$$g\left(Y_{\text{mix}}\right) = \beta_0 + \beta^* d + \epsilon \tag{3.32}$$

on the total dose is fitted to the mixture along each ray. Then, the null hypothesis of additivity can be formulated as

$$H_0 : \beta_l^* = \sum_{i=1}^{k} a_{i(l)} \beta_i ; \, l = 1, \ldots, L \tag{3.33}$$

where β_l^* is the slope of the regression of the mixture along the *l*th ray and the same background parameter β_0 is used for the mixture and each individual chemical. Note also that in fitting the model (3.24) to individual chemicals and model (3.32) to the mixture, only linear terms are considered. If it is desired to include higher-order terms or different intercepts, that is also possible, but more parameters would have to be estimated from the data. Following Gennings et al. (1997), Casey et al. (2004) use a threshold additivity model to express the dose-response relationship. Although biologically that may be more plausible, but mathematically that model simply adds a parameter for the threshold along each ray and for simplicity, here we use a non-threshold model and refer the reader to Casey et al. (2004) for more details where a Wald-type test is proposed for testing (3.31). Now, if we let

$$\theta = \left(\beta_0, \beta_1, \ldots, \beta_k, \beta_1^*, \beta_2^*, \ldots, \beta_L^*\right)^T$$

be the vector of all the model parameters, then the statistic

$$T(\hat{\theta}) = \frac{(B\hat{\theta})^T (BSB^T)^{-1} (B\hat{\theta})}{L} \tag{3.34}$$

has a χ^2 distribution with L degrees of freedom. In (3.34), S is the variance–covariance matrix of θ, and B is an appropriate $L \times (k+L+1)$ matrix defining the contrasts to be tested. Specifically, for the additivity hypothesis,

$$B = \begin{pmatrix} 0 & -a_{1(1)} & -a_{2(1)} & \cdots & -a_{k(1)} & 1 & 0 & 0 \\ 0 & -a_{1(1)} & -a_{2(2)} & & -a_{k(2)} & 0 & 1 \cdots 0 \\ & \vdots & & \ddots & & \vdots & \\ 0 & -a_{1(L)} & -a_{2(L)} & \cdots & -a_{k(L)} & 0 & 0\ldots 1 \end{pmatrix}.$$

Using (3.34), confidence intervals of the desired significance level can also be constructed.

Example 3.5

In this example, we use the summary data given in Gennings et al. (2002) presented here in Table 3.7. The experiment was conducted to assess the cytotoxic interaction of four metals arsenic, cadmium, chromium, and lead, all four known to be dangerous and toxic in humans. Cells from normal human epidermal keratinocytes were cultured and treated with

TABLE 3.7

Cytotoxicity in Human Epidermal Keratinocytes

Chemical	Dose	Mean	Variance	Sample Size
Arsenic	0	100	45.7	9
	0.3	108.9	146.6	9
	1	99.6	78.1	9
	3	84.1	279	9
	10	45.1	268	9
	30	8.4	108	9
Chromium	0	100	137	6
	0.3	105.1	119	9
	1	99.7	124	9
	3	79.8	304	9
	10	23.9	446	9
	30	6.6	78	9
Cadmium	0	100	27	6
	3	92.6	130	9
	10	51.9	1236	9
	30	20.2	829	9
	100	3.8	23	9
	300	4.6	28	9
Lead	0	100	8.4	6
	3	95.5	79	9
	10	91	63	9
	30	98.8	236	9
	100	93.7	1838	9
	300	28.9	297	9

Data from Gennings et al. (2002).

increasing doses of the four metals individually as well as a mixture of the metals along a single ray ($L=1$). The mixture point of interest contained concentrations of the metals to produce 50% lethality (LD_{50}). The concentrations of the four metals in the mixture were 7.7 µM, 4.9 µM, 6.1 µM, and 100 µM with a total dose of 118.7 µM for arsenic, chromium, cadmium, and lead respectively and the mixture was diluted to produce seven total doses. Thus

$$a_{\text{arsenic}(1)} = \frac{7.7}{118.7} = 0.0649, \quad a_{\text{chromium}(1)} = \frac{4.9}{118.7} = 0.0413,$$

$$a_{\text{cadmium}(1)} = \frac{6.1}{118.7} = 0.0514, \quad a_{\text{lead}(1)} = \frac{100}{118.7} = 0.8425.$$

The summary data for each chemical and the mixture are given in Tables 3.7 and 3.8. Once again, for the purpose of illustration, we simulate representative data sets from this experiment; see Appendix 2 for details. A linear model, i.e., the identity link function, was used to fit the individual chemical data as well as the mixture data. Every term in the individual chemical data was significant as well as the overall model. Table 3.9 gives the parameter estimates along with their associated p-values.

Since $L=1$, for testing additivity,

$$B = \begin{pmatrix} 0 & -0.0649 & -0.0413 & -0.0514 & -0.8425 & 1 \end{pmatrix}^{T}$$

and we find that $BSB^{T} = 0.005384$, $B\hat{\theta} = 0.51337$, and from (3.30) we get $T(\hat{\theta}) = 49.95$ and the hypothesis of additivity is strongly rejected.

3.3.7 Test of Additivity When a Single Chemical Data Are Not Available

The procedures described in the last sections for testing additivity, either for full factorial design mixtures or for fixed-ratio ray design mixtures, rely

TABLE 3.8

Cytotoxicity of a Mixture of Four Metals

Mixture Dilution	Total Concentration	Mean	Variance	Sample Size
0.0014	0.16	116.6	315	9
0.004	0.48	95.7	1.9	9
0.0123	1.46	96	216	9
0.037	4.39	76.5	92	9
0.111	13.2	64.3	14	9
0.333	39.6	50.1	262	9
1	118.7	31.1	207	9

Data from Gennings et al. (2002).

TABLE 3.9

Parameter Estimates for Individual and Mixture of Four Metals

Parameter	Estimate	S.E.	p-Value
β_0 (Additivity)	90.0097	2.3437	$<2\times10^{-16}$
β_1 (Arsenic)	−2.7805	0.3056	$<2\times10^{-16}$
β_2 (Chromium)	−3.2810	0.3056	$<2\times10^{-16}$
β_3 (Cadmium)	−0.3600	0.0306	$<2\times10^{-16}$
β_4 (Lead)	−0.2124	0.0306	$<4.87\times10^{-11}$
β_0 (Mixture)	88.0192	2.9722	$<2\times10^{-16}$
β_1^* (Mixture)	−0.4983	0.0624	$<4.79\times10^{-11}$

on availability of data for single chemicals in the mixture. Such data may not always be available or may be very costly to obtain, especially when the number of chemicals in the mixture is relatively large. Another issue that might affect the results when the number of chemicals in the mixture is relatively large is the design of the experiment. Change of environment, different batches of raw material, and other similar factors can affect the outcome and cause a shift in the dose response. In this section, we present a method for assessing additivity and departure from additivity in mixtures in the absence of data for each component of the mixture. We only consider fixed-ratio ray mixtures due to their efficiency in experimental design.

In Section 3.1, it was shown that two chemicals can be declared as additive when the coefficient of the interaction term β_{12} is zero in the generalized linear model

$$g(Y_{\mathrm{mix}}) = \beta_0 + \beta_1 d_1 + \beta_2 d_2 + \beta_{12}d_1 \times d_2 + \epsilon \qquad (3.35)$$

where Y_{mix} is the response of interest, $g(.)$ is the link function, and ϵ is the random error. Thus, testing the additivity of the chemicals is equivalent to testing the null hypothesis $H_0 : \beta_{12} = 0$ and significance of the test would imply departure from additivity. Following Carter et al. (1988) and Meadows et al. (2002), this methodology for testing additivity can be extended to more than two chemicals. Using the additivity index (3.23), analogous to (3.35) we find that for a mixture of k chemicals in the response surface

$$g(Y_{\mathrm{mix}}) = \beta_0 + \sum_{i=1}^{k}\beta_i d_i + \sum_{\substack{i=1 \\ i<j}}^{k}\sum_{j=1}^{k}\beta_{ij}d_i d_j$$

$$+ \sum_{\substack{i=1 \\ i<j<l}}^{k}\sum_{j=1}^{k}\sum_{l=1}^{k}\beta_{ijl}d_i d_j d_l + \cdots + \beta_{123\ldots k}d_1 d_2 \ldots d_k + \epsilon \qquad (3.36)$$

the additivity amounts to all the cross product term vanishing; i.e. all the coefficients for the cross product terms are equal to 0 and significance of the corresponding hypothesis indicates departure from additivity. However, for the case of fixed-ratio ray mixtures with mixing ratios of $a_1, a_2, \ldots, a_k$ for a total dose d of the mixture, we can write (3.36) as

$$g(Y_{\text{mix}}) = \beta_0 + \sum_{i=1}^{k} a_i \beta_i d + \sum_{\substack{i=1 \\ i<j}}^{k} \sum_{j=1}^{k} a_i a_j \beta_{ij} d^2$$

$$+ \sum_{\substack{i=1 \\ i<j<l}}^{k} \sum_{j=1}^{k} \sum_{l=1}^{k} a_i a_j a_l \beta_{ijl} d^3 + \cdots + a_1 a_2 \ldots a_k \, \beta_{123\ldots k} d^k + \epsilon$$

(3.37)

leading to a polynomial of degree k in d,

$$g\left(Y_{\text{mix}}\right) = \sum_{i=0}^{k} \beta_i^* d^k + \epsilon$$

(3.38)

where

$$\beta_0^* = \beta_0$$

$$\beta_1^* = \sum_{i=1}^{k} a_i \beta_i$$

$$\beta_2^* = \sum_{i=1}^{k} \sum_{j=1}^{k} a_i a_j \beta_{ij}$$

$$i < j$$

$$\beta_3^* = \sum_{i=1}^{k} \sum_{j=1}^{k} \sum_{l=1}^{k} a_i a_j a_l \, \beta_{ijl}$$

$$i < j < l$$

$$\vdots$$

$$\beta_k^* = a_1 a_2 \ldots a_k \, \beta_{123\ldots k}.$$

Thus it follows that other than the linear term in (3.37) which represents additivity, each subsequent term represents an interaction among the components of the mixture. The quadratic term represents interaction among pairs of chemicals, the third-degree term represents interaction among any three chemicals, and so on. Hence, we can deduce that components of the chemical can be assumed to be additive if the coefficients in the polynomial (3.38) after the linear term are all equal to 0. In other words, evidence of non-linearity and curvature indicates nonadditivity. Therefore, the null hypothesis of additivity can be formulated as

$$H_0 : \beta_2^* = \beta_3^* = \cdots = \beta_k^0 = 0.$$

To test this hypothesis, as described in Meadows et al. (2002), if it can be assumed that the transformed responses $g(Y_{mix})$ have a normal distribution, one can simply use the linear regression of the full model

$$Y_{mix} = \beta_0^* + \beta_1^* d_1^* + \beta_2^* d_2^* + \cdots + \beta_k^* d_k^* + \epsilon$$

where $d_i^* = d^i\ i = 1,\ldots,k$ and linear regression of the reduced model

$$Y_{mix} = \beta_0^* + \beta_1^* d_1^* + \epsilon$$

and perform a partial F-test in the usual way by using the residual sum of squares R_{full} of the full model and the residual sum of squares $R_{reduced}$ of the reduced model and calculating

$$f = \frac{(R_{reduced} - R_{full}) \big/ (k-1)}{R_{full} \big/ (N-k-1)} \qquad (3.39)$$

where N is the total number of measurements from the mixture. This statistic has an F distribution with $(k - 1)$ and $(L - k - 1)$ degrees of freedom and the null hypothesis of additivity is rejected when the value of F in (3.39) exceeds that of the F distribution at the desired level of significance.

Casey et al. (2004), give a more general Wald-type version of the test in (3.39) similar to (3.34) where any linear contrast may be tested. They also consider a threshold model for the dose response. As commented before, a threshold model may biologically be more plausible, but mathematically simply adds more parameters to the model. But the nature of the testing of the additivity hypothesis remains the same. Note also that the methodology described here can be used to examine the effect of a subset of chemicals. The method is similar to testing the effect of a subset of predictors by using the partial F-test in linear regression. For more on this topic, we refer to Kleinbaum et al. (2014).

Example 3.6

In the last example, we used a methodology based on using data from each chemical component in the mixture as well as the data from the mixture to test the additivity hypothesis. In this example we use only

TABLE 3.10

Parameter Estimates for the Polynomial Regression of Mixture of Four Metals

Parameter	Estimate	Standard Error	p-Value
β_0^*	107.4	3.325	$<2 \times 10^{-16}$
β_1^*	−10.12	2.264	$<4 \times 10^{-5}$
β_2^*	0.735	0.236	0.00286
β_3^*	−0.0169	0.006	0.00635
β_4^*	0.000096	0.000035	0.00791

the data from the mixture experiment with a total of 63 measurements to examine departure from additivity. Using, once again, the identity function as the link and applying the *lm* function in R to fit a polynomial regression of degree 4, we find that the overall model is highly significant, $p < 2.2 \times 10^{-16}$, and that the coefficient of each term in the model is also significant. Table 3.10 gives the parameter estimates along with their standard errors and the associated p-values. Moreover, we find that for the linear model $R_{reduced} = 24257.55$ with 61 degrees of freedom while from the polynomial model we have $R_{full} = 1025.76$ with 58 degrees of freedom. Thus from (3.39) we have

$$f = \frac{(24257.55 - 1025.76) \big/ 3}{1025.76 \big/ 58} = 437.87$$

which, compared with the critical value of the F distribution with 3 and 58 degrees of freedom, clearly rejects additivity.

3.4 Additional Topics

• **Power and Sample Size**

An important issue in designing experiments for testing additivity of chemical components in a mixture is the question of power and sample size. The traditional approach to detecting and characterizing departure from additivity and existence of interaction is the response surface methodology, supported by factorial designs. For such designs, it is possible to detect departure from additivity at any specific mixture point by comparing the observed response with that predicted under additivity. Meadows-Shropshire et al. (2005) discuss the question of power and sample size for such designs. However, we have seen that the classical approach of factorial design becomes prohibitive and highly inefficient as the number of chemicals in the mixture rises. Since humans are generally exposed to a multitude of chemicals through different routes of exposure, the fixed-ratio ray designs provide a much more attractive and cost-effective method for characterizing and detecting possible departure from additivity. The power and sample size determination for such designs are addressed in Casey et al. (2006).

- **Sufficient Similarity**

  A problem that has attracted the attention of many researchers in recent years is the question of sufficient similarity in chemical mixtures. Although it is always preferred to have dose-response studies on whole mixtures for the purpose of risk assessment, the fact is that such data are seldom available especially in the case of complex mixtures where usually a large number of chemicals may be involved. When the toxicity assessment data for a chemical mixture are not available, the guidelines developed by the US EPA (1986, 2000) allow the consideration of available toxicity data for a surrogate mixture that is thought to be similar to the mixture under consideration. So the question that arises is when we can call two mixtures "similar." Clearly, as explained by Lipscomb et al. (2010), it would be reasonable to expect the "surrogate mixture" to have similar components as compared with the mixture under study and that the components are mixed in similar proportions. So, how do we make a judgment regarding similarity? The EPA guidelines rely on expert judgment to define sufficient similarity in terms of biological processes and endpoints. However, Stork et al. (2008) consider this problem in fixed-ratio ray mixtures from a statistical viewpoint. Assuming that dose-response data are available from a so-called "reference" mixture, using a mixed model, they determine a region of dose-response similarity based on the ideas of statistical equivalence of hypotheses testing (Wellek, 2003). As pointed out in Stork et al. (2008), the biological basis for similarity is of course critical in determining sufficient similarity, but since similarity is a relative term, that could pose some potential problem. The statistical approach creates a foundation and provides a structure for a better understanding of the underlying biological concepts. There are still several potential problems that require attention of statisticians, and the subject of sufficient similarity is still under consideration by many researchers. For the most recent problems and developments on this topic we refer to Catlin et al. (2018).

Exercises

1. In this exercise, you mimic Example 1 of this chapter for a different data set. Kodell and Pounds (1991) describe an experiment with

TABLE 3.11

Proportion of Lactate Dehydrogenase (LDH) Release in Mice

Mercury Chloride		Cadmium Chloride		Mixture*	
Dose (μM)	LDH	Dose (μM)	LDH	Dose (μM)	LDH
20	0.770	4	1.000	14.7	0.576
	0.496		0.972		0.537
	0.486		0.954		0.434
15	0.589	3	0.914	11.0	0.351
	0.563		0.968		0.439
	0.605		0.967		0.387
12.5	0.273	2.5	0.823	9.1	0.264
	0.268		0.753		0.177
	0.280		0.817		0.260
10	0.145	2.0	0.327	7.3	0.179
	0.201		0.366		0.172
	0.201		0.427		0.135
8.3	0.122	1.7	0.326	6.1	0.147
	0.105		0.495		0.080
	0.203		0.396		0.064
6.7	0.139	1.3	0.159	4.9	0.106
	0.097		0.200		0.111
	0.099		0.286		0.102

*The mixture contains mercury chloride and cadmium chloride in the ratio of 10:1.
Data from Kodell and Pounds (1991).

model hepatotoxicants and cultured rat hepatocytes to demonstrate the application of statistical methods in predicting the toxicity of a mixture of two chemicals. Primary cell cultures of adult rat hepatocytes were prepared from male Sprague-Dawley rats and maintained as described in Pounds et al. (1982). After adaptation to culture conditions, cultures were treated in triplicates with selected concentration of cadmium chloride, a known human carcinogen, and mercury chloride, known to be highly toxic in humans. Exposure to concentrations of a single or a mixture of the two chemicals was for 20 hours. The release to a soluble enzyme lactate dehydrogenase was measured as the fraction of the total activity for each culture. The data are reproduced in Table 3.11. Since the data are expressed as proportions, we may assume that it is probabilistic in scale. Kodell and Pounds (1991) use the angle transformation to linearize the data and demonstrate their methodology of using the dose or the log-dose to model the joint toxicity. Razzaghi and Kodell (1992) use the same transformation to show that the Box–Cox model can provide a better characterization of the toxicity of the mixture.

   a. Use the logistic or the probit link as the linearizing transformation to fit the data for each chemical and the mixture.

   b. Express the risk prediction model for the mixture under both concentration additivity and response additivity assumptions.

   c. Now, use the Box–Cox model to derive the most appropriate power transformation for all models obtained in (a.) and compare the results.

2. Use the data from Exercise 1 to test whether or not the two chemicals, mercury chloride and cadmium chloride, are interactive and determine the type of interaction.

3. Generalize expression (3.17). Suppose in a binary mixture, the ratio of the two chemicals is a:b. Derive an expression for the sum of the toxic units analogous to (3.17) and give an interpretation.

4. Use Table 3.5 to generate a representative sample for the mixture effect of the 5 PAH mixture study where the response is the number of tumors by applying a methodology similar to that described in Appendix 1. Use the simulated data to fit the regression model for a 2^5 factorial experiment and test for significance of main effects and all interaction terms.

5. In this exercise, you are required to mimic Example 3.5 for a different data set. Gennings et al. (1997) and Meadows et al. (2004) describe an experiment for the study of the effect of a mixture of four trialomethanes formed as disinfection by-products (DBPs) for chlorinating water. The four chemicals used in the study were bromodichloromethane (BDCM), chlorodibromomethane, (CDBM), chloroform ($CHCl_3$), and bromoform ($CHBr_3$). The purpose of the study was to determine if a mixture of these chemicals resulted in a response based on additivity or if there was a departure from additivity due to the interaction of these chemicals. Female CD-1 mice were exposed to each individual chemical or a mixture at a fixed ratio of 0.34:0.29:0.32:0.05 for BDCM:CDBM:$CHCl_3$:$CHBr_3$. There was a total of seven experiments, four for each chemical and three for chemical mixtures at total doses of 0.0 mg/kg, 1.5 mg/kg, and 3.0 mg/kg. The response of interest was the liver damage, as measured by serum sorbitol dehydrogenase (SDH) levels. Tables 3.12 and 3.13 give the summary results for the individual chemical experiments and the mixture studies respectively.

   a. Simulate representative data sets for the individual chemical and the mixture experiments using the methodology described in Appendix 2.

   b. Perform a test of additivity using representative data sets generated in a. analogous to Example 3.5.

TABLE 3.12

Liver Damage in CD-1 Mice

Chemical	Dose	Mean	Standard Deviation	Sample Size
Bromodichloromethane (BDCM)	0	19.3	3.77	10
	0.152	21.4	3.44	10
	0.305	26.6	7.97	11
	0.76	39.6	12.4	11
	1.52	154.4	133.8	10
	3.05	187.4	163.9	8
Chlorodibromomethane (CDBM)	0	21.2	3.07	13
	0.152	25.6	5.09	16
	0.305	32	9.76	18
	0.76	167.9	112.1	14
	1.52	146.7	123.9	5
	3.05*	–	–	–
Chloroform (CHCl$_3$)	0	15.6	3.27	12
	0.152	16.8	4.2	13
	0.305	21.3	5.85	15
	0.76	30.3	19.9	15
	1.52	50.8	22.5	13
	3.05	80.2	9.35	12
Bromoform (CHBr$_3$)	0	21.3	5.97	16
	0.152	24.1	6.77	14
	0.305	23.8	4.14	15
	0.76	38.2	24.1	16
	1.52	87.6	21.8	10
	3.05	130	125	14

*No mice survived at this dose level.
Data from Gennings et al. (1997).

TABLE 3.13

Liver Damage in CD-1 Mice Exposed to a Mixture of Disinfection By-Products (DBPs)

Mixture Group	Total Concentration	Mean Serum Sorbitol Dehydrogenase (SDH)	Standard Deviation (SDH)	Sample Size
Control	0	17.42	3.43	27
Chlorination*	0.05	19.61	4.36	17
Chlorination*	1.5	46.29	19.26	17
Chlorination*	3	76.22	12.1	16

*Mixing ratio of (0.34:0.29:0.32:0.05).
Data from Meadows et al. (2004).

6. In Exercise 5, assume that no single chemical data are available. Use only the representative data simulated for the mixture study in Exercise 5, to perform a test of additivity similar to Example 3.6.

References

Barenbaum, M. C. (1985). The expected effect of combination of agents: The general solution. *Journal of Theoretical Biology*, 114, 413–31.

Barenbaum, M. C. (1989). What is synergy? *Pharmacological Review*, 41, 93–141.

Carter, W. H. Jr., Gennings, C., Stainwalis, J. G., Campbell, E. D., and White, K. L. Jr. (1988). A statistical approach to the construction and analysis of isobologram. *Journal of American College of Toxicology*, 7, 963–73.

Casey, M., Gennings, C., Carter, W. H. Jr., Moser, V., and Simmons, J. E. (2004). Detecting interaction and assessing the impact component subset in a chemical mixture using fixed-ratio ray designs. *Journal of Agricultural, Biological, and Environmental Statistics*, 9, 339–61.

Casey, M., Gennings, C., Carter, W. H. Jr., Moser, V., and Simmons, J. E. (2006). Power and sample size determination for testing the effect of subsets of compounds on mixtures along fixed-ratio rays. *Journal of Ecological and Environmental Statistics*, 13, 11–23.

Catlin, N. R., Collins, B. J., Auerbach, S. S., Ferguson, S. S., Harnly, J. M., Gennings, C., Waidyanatha, S., Rice, G. E., Smith-Roe, S. L., Witt, K. L., and Rider, C. V. (2018). How similar is similar enough? A sufficient similarity case study with *Ginkgo biloba* extract. *Food and Chemical Toxicology*, 118, 328–39.

Fraser, T. R. (1872). An experimental research on the antagonism between the actions of physostigma and antropia. *Proceedings of the Royal Society of Edinburgh*, 7, 506–11.

Gennings, C. (1995). An efficient experimental design for detecting departure from additivity in mixtures of many chemicals. *Toxicology*, 10, 189–97.

Gennings, C. and Carter, W. H. Jr. (1995). Utilizing concentration-response data from individual components to detect statistically significant departures from additivity in chemical mixtures. *Biometrics*, 51, 1264–77.

Gennings, C., Schwartz, P., Carter, W. H. Jr., and Simmons, J. E. (1997). Detection of departures from additivity in mixtures of many chemicals with a threshold model. *Journal of Agricultural, Biological, and Environmental Statistics*, 2, 198–211.

Gennings, C., Carter, W. H. Jr., Campain, J. A., Bae, D., and Yang, R. S. H. (2002). Statistical analysis of cytotoxicity in human epidermal keratinocytes following exposure to a mixture of four metals. *Journal of Agricultural, Biological, and Environmental Statistics*, 7, 58–73.

Gessner, P. K. and Cabana, B. E. (1970). A study of the interaction of the hypnotic effects and of the toxic effects of chloral hydrate and ethanol. *Journal of Pharmacology and Experimental Therapeutics*, 174, 247–59.

Hewlett, P. S. (1969). Measurement of potency of drug mixtures. *Biometrics*, 25, 477–87.

Hochberg, Y. (1990). A sharper Bonferroni procedure for multiple tests of significance. *Biometrika*, 75, 800–2.

Kleinbaum, D. G., Kupper, L. L., Nizam, A., and Rosenberg, E. (2014). *Applied regression analysis and other multivariate methods,* 5th edition. Cengae Learning, Boston.

Kodell, R. L. and Pounds, J. G. (1991). Assessing the toxicity of mixtures of chemicals. In *Statistics and toxicology,* D. Krewski and C. Franklin eds. Gordon Breach, New York, NY.

Lipscomb, J. C., Lambert, J. C., and Teuscher, L. K. (2010). Chemical mixtures and cumulative risk assessment. In *Principles and practice of mixture toxicology,* M. Mumtaz ed. Wiley-VCH Verlag, Germany.

Loewe, S. (1953). The problem of synergism and antagonism of combined drugs. *Areheimittle Forshung,* 3, 285–90.

Loewe, S. and Muischnek, H. (1926). Uber kombination-wikugen. I. mitteilung: Hilfsmittle der fragestellung. *Archives of Experimental Pathology and Pharmacology,* 114, 313–26.

McCullagh, P. and Nelder, J. A. (1989). *Generalized linear models.* Chapman and Hall, London.

Meadows, S. L., Gennings, C., Carter, W. H. Jr., and Bae, D. S. (2002). Experimental design for mixtures of chemicals along fixed-ratio rays. *Environmental Health Perspectives,* 110, 979–83.

Meadows-Shropshire, S. L., Gennings, C., Carter, W. H., and Simmons, J. E. (2004). Analysis of mixtures of drugs/chemicals along a fixed-ratio rays without single-chemical data to support an additivity model. *Journal of Agricultural, Biological, and Environmental Statistics,* 9, 500–514.

Meadows-Shropshire, S. L., Gennings, C., and Carter, W. H. Jr. (2005). Sample size and power determination for detecting interaction in mixtures of chemicals. *Journal of Agricultural, Biological, and Environmental Statistics,* 10, 104–17.

Miller, R. G. Jr. (1991). Simultaneous statistical inference. Springer-Verlag, New York.

Montgomery, D. C. (2017). *Design and analysis of experiments,* 9th edition. Wiley, New York, NY.

Nesnow, S., Mass, M. J., Ross, J. A., Galati, A. J., Lambert, G. R., Gennings, C., Carter, W. H. Jr., and Stoner, G. D. (1998). Lung tumorigenic interactions in strain A/J mice of five environmental polycyclic aromatic hydrocarbons. *Environmental Health Perspectives,* 106, 1337–46.

Pounds, J. G., Wright, R., and Kodell, R. L. (1982). Cellular metabolism of lead: A kinetic analysis in the isolated rat hepatocyte. *Toxicology and Applied Pharmacology,* 66, 88–101.

Razzaghi, M. and Kodell, R. L. (1992). Box-cox transformation in the analysis of combined effect of mixtures of chemicals. *Environmetrics,* 3, 319–34.

Shelton, D. W. and Weber, L. J. (1981). Quantification of the joint effects of mixtures of hepatotoxic agents: Evaluation of theoretical model in mice. *Environmental Research,* 26, 33–41.

Sorensen, H., Cedergreen, N., Skovgraad, I. M., and Streibig, J. C. (2007). An isobole-based statistical model and test for synergism/antagonism in binary mixture toxicity experiments. *Journal of Environmental and Ecological Statistics,* 14, 383–97.

Stork, L. G., Gennings, C., Carter, W. H. Jr., Teuscheler, L. K., and Carney, E. W. (2008). Empirical evaluation of sufficient sufficiency in dose-response for environmental risk assessment of chemical mixtures. *Journal of Agricultural, Biological, and Environmental Statistics,* 13, 313–33.

Streibig, J. C., Kudsk, P., and Jensen, J. E. (1998). A general joint action model for herbicide mixtures. *Pesticide Science,* 53, 21–8.

US Environmental Protection Agency (1986). Guidelines for the health risk assessment of chemical mixtures. EPA/630/R-98/002.

US Environmental Protection Agency (2000). Supplementary guidance for conducting health risk assessment of chemical mixtures. EPA/630/R-00/002.

Volund, A. (1992). Dose-response surface bioassay. In XVIth International Biometric Conference, Vol. II, P. 249, Hamilton, New Zealand.

Wellek, S. (2003). *Testing statistical hypothesis of equivalence*. Chapman and Hall CRC, Boca Raton, FL.

Wu, C. F. J. (1986). Jackknife, bootstrap, and other resampling methods in regression analysis (with discussion). *The Annals of Statistics*, 14, 1261–350.

4

Models for Carcinogenesis

Mathematical modeling of cancer dose-response for the purpose of individualized risk estimation dates back several decades. With limited amount of data, several attempts were made to use basic statistical models to express the process of carcinogenesis. As more biological information became available, along came more sophisticated statistical tools leading to more advanced and more realistic models with sophisticated statistical tools. The ultimate goal of these models has been to assess the lifetime chance of developing cancer in humans. Since humans are exposed to a multitude of chemicals at various levels with irregular exposure intervals, estimation of the probability of incidence is inherently a difficult problem. The major goal of toxicological bioassay experiments with animals has been to try and understand the mechanism of carcinogenesis. However, early attempts at modeling cancer dose-response did not incorporate any information about the carcinogenic mechanism of action. These models are termed the Tolerance Distribution Models. The models were used to simply fit the available data in order to express the relationship between exposure and incidence rate with no understanding of the underlying mechanism. It was soon realized that biologically based models that incorporate the specific biological parameters relating to the cancer mechanism have a better chance of success and are more likely to produce more reliable risk estimates. The so called mechanistic models were developed for probability of cancer incidence based on the theoretical understanding of the carcinogenesis process. Since it became known that the transformation of normal cells into cancer cells is a multistage process in which the stem cells go through a series of genetic changes that render the cells malignant, the multistage model was developed. A major breakthrough in the development of cancer dose-response analysis occurred when the pharmacokinetic information was considered additionally. These models were known as Physiologically Based Pharmacokinetic (PBPK) models. As described by Dewanji (2008), the dose-response shape is not only determined by the characteristics of the carcinogenic mechanism taking place at the target tissue, but also the role of the pharmacokinetic process in the distribution and absorption of the chemical. In this chapter, we demonstrate the process of the progression of models by starting with the simple dose-response models and finish by describing the features and properties of the PBPK model.

Cancer risk is measured by hazard rate or age-specific incidence rate. The hazard function is defined as the instantaneous rate of change in the probability of cancer incidence. Thus if T is the response time and $P(t) = P(T \leq t)$ is the probability that cancer occurs by time t, the hazard is defined as

$$h(t) = \lim_{\Delta t \to 0} \frac{P(t \leq T \leq t + \Delta t \mid T \geq t)}{\Delta t} \qquad (4.1)$$

which is the theoretical representation of cancer incidence or cancer death rate. Thus in order to estimate the hazard, the number of cases is divided by the number of individuals at risk. Using the above definition, we have for the incidence probability up to time t,

$$P(t) = 1 - \exp\left\{ -\int_0^t h(x)\,dx \right\} \qquad (4.2)$$

and thus the lifetime risk of developing cancer is given by

$$R = 1 - \exp\left\{ -\int_0^L h(x)\,dx \right\} \qquad (4.3)$$

where L is the individual's lifetime. When incidence rate is small, $P(t)$ can be approximated by $\int_0^t h(x)\,dx$, i.e. the cumulative incidence function. Therefore cancer models consider modeling $P(t)$ or $h(t)$.

4.1 Tolerance Distribution Models

Tolerance distribution models simply express the distribution of individual toxic thresholds in the population. It is assumed that each individual (animal in carcinogenicity experiments) in the population has a tolerance for the carcinogenic chemical. Once the dose exceeds this tolerance limit, the individual develops tumor. The level of tolerance varies in individuals and variation follows a statistical distribution. Thus tolerance distribution models are simply parameterized statistical distributions. Several standard distributions have been used for modeling the probability of cancer as a function of dose. Note that in tolerance distribution models, time is not a variable and thus given the same dose, the probability of developing cancer by a certain time is constant for all individuals. Thus these models are homogeneous with respect to time. The most common tolerance distribution models are:

1. The log-probit model which assumes a lognormal distribution for the tolerance. Thus it is assumed that the logarithm of variation in tolerance has a normal distribution with mean μ and standard deviation of σ. Therefore the probability of developing a tumor at the exposure level of d is the probability that the tolerance is below d and is given by

$$P(d) = \Phi\left(\frac{\log d - \mu}{\sigma}\right) = \int_{-\infty}^{(\log d - \mu)/\sigma} (2\pi)^{-1/2} \exp(-z^2/2)\,dz. \qquad (4.4)$$

The log-probit model is probably the most popular tolerance distribution model and has proved to provide adequate fit to many toxicological data. An important feature of this model is that it states that there are relatively few extremely sensitive or extremely resistant animals in the population.

2. The log-logistic model where it is assumed that the tolerance has a logistic distribution. The probability of tumor for exposure level of d in this case is given by

$$P(d) = \frac{\exp(\alpha + \beta \log d)}{1 + \exp(\alpha + \beta \log d)}. \qquad (4.5)$$

The advantage of the logit model is that it allows fitting data with a higher degree of variation.

3. The Weibull model for which the probability of developing tumor has a Weibull distribution given by

$$P(d) = 1 - \exp(-\beta d^{\alpha}). \qquad (4.6)$$

This model generally provides a more conservative estimate of cancer risk.

There are also other tolerance distribution models, but since these models are known to be too simplistic to describe the complex process of carcinogenesis, we do not discuss them any further.

Example 4.1

Quast et al. (1980) describe an experiment to study the carcinogenic effects of acrylonitrile. Sprague-Dawley rats were exposed to saline (control) and three non-zero doses of the chemical through drinking water and number of rats developing tumor was noted. Table 4.1 gives the dose-response data. Fjeld et al. (2006) use this data set to show the

TABLE 4.1

Carcinogenic Effects of Scrylonitrile in Rats

Daily Dose mg/kg	Number Exposed	Number with Tumor	Proportion
0	80	4	0.0500
3.42	47	18	0.3830
8.53	48	36	0.7500
21.28	48	45	0.9375

Data from Quast et al. (1980).

TABLE 4.2

Parameter Estimates for Log-Logistic Model

	Estimate	Std. Error	t-value	p-value
α	4.509	0.309	14.567	0.005
β	1.732	0.221	7.834	0.016

application of LOAEL and NOAEL. Here, for illustration, we fit the log-logistic model as the tolerance distribution. Using the function *drm* (dose-response model) in the *drc* package in R, the parameter estimates can readily be obtained. Table 4.2 gives the parameter estimates along with their associated standard errors and the p-values. Figure 4.1 displays the experimental points with the fitted dose-response model. Although the parameter estimates are significant, clearly the model appears to grossly overestimate the response specially at high doses.

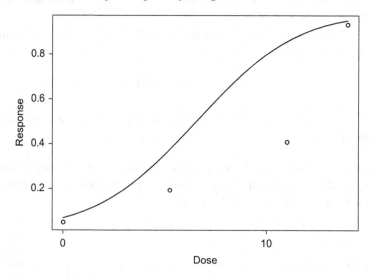

FIGURE 4.1

Fitted log-logistic model to the dose-response data for the carcinogenicity of acrylonitrile in rats.

4.2 Mechanistic Models

In mechanistic models the theoretical understanding of the biology of cancer is utilized to predict the probability of incidence. A mechanistic model is based on an understanding of the behavior of a system's components. Thus mechanistic models assume that a complex system can be analyzed by examining its individual components. In a mechanistic model, a relationship is hypothesized between the variable and the data in a way that the biological process is thought to have given rise to the data. Therefore, in a mechanistic model, parameters of the model all have biological interpretation and can be measured independently of the data. Below is the list of popular mechanistic models.

4.2.1 One-Hit Model

The oldest mechanistic model of carcinogenesis is the one-hit model. The one-hit model assumes that cancer is the result of a single event (hit) in a single cell. Thus only one genetic change is sufficient to transfer a normal cell into a cancer cell and cancer risk is present at any dose, i.e. no threshold. The distribution for the one-hit model is the cumulative distribution function of the exponential distribution and is given by

$$P(d) = 1 - e^{-\beta d} \quad \beta > 0. \tag{4.7}$$

Note that this model has only one parameter and indicates that there is zero risk of cancer at zero dose. Also, since

$$P(d) = \beta d - \frac{(\beta d)^2}{2!} + \frac{(\beta d)^3}{3!} + \cdots, \tag{4.8}$$

the one-hit model is essentially linear in the low- and middle-dose levels and is very conservative. Thus the model can be used to fit a single dose point and is generally not appropriate for fitting a data set with more than one dose since the linearity is most likely not satisfied. The one-hit model is considered to be a very simplistic representation of the biological process of carcinogenesis and it is highly unlikely that any single molecule would be biologically capable of causing cancer. A more general form of the one-hit model is expressed as

$$P(d) = 1 - e^{-(\beta_0 + \beta_1 d)} \quad \beta_0, \beta_1 > 0$$

which incorporates a non-zero probability of cancer for the background zero dose.

4.2.2 Multi-Hit (*k*-Hit) Model

Carcinogenesis is known to be a multistep process, with several genetic transformations required to produce a tumor cell. The multi-hit model is based on
the assumption that a fixed number of identical events (hits) is required to
produce a cancer cell. The multi-hit model is defined as the cumulative probability of a gamma distribution given by

$$P(d) = \int_0^d \frac{\beta^k t^{k-1} e^{-\beta t}}{(k-1)!} dt = \int_0^{\beta d} \frac{t^{k-1} e^{-t}}{(k-1)!} dt \quad \beta > 0 \tag{4.9}$$

where k is the least number of hits required to produce the cancer cell. This
number also determines the shape of the model. Clearly the higher values of
k lead to lower probability of cancer at low doses, but that probability rises
sharply as the dose increases. This is a two-parameter model and tends to
give a better estimate of the risk than the one-hit model. In the most general
case, k may be assumed to be a non-integer real number estimated from the
data. Using the relationship between Poisson and gamma distributions, the
probability that exactly k hits would be required to induce cancer as a function of dose would become

$$P(d) = 1 - \sum_{r=0}^{k-1} \frac{(\beta d)^r e^{-\beta d}}{r!}. \tag{4.10}$$

Similar to the one-hit model, the more general expression of the multi-hit
model incorporating a background probability is given by

$$P(d) = \int_0^{(\beta_0 + \beta_1 d)} \frac{t^{k-1} e^{-t}}{(k-1)!} dt \quad \beta_0, \beta_1 > 0. \tag{4.11}$$

4.2.3 Weibull Model

The Weibull model is defined as

$$P(d) = 1 - e^{-\beta d^k} \quad \beta, k > 0 \tag{4.12}$$

which is another generalization of the one-hit model. In fact for $k = 1$, the one-
hit (linear) model is resulted. When $k < 1$, the model is concave (sub-linear)
and when $k > 1$, the model is convex (supra-linear). The risk estimates derived
using the Weibull model tend to lie between the estimates from the one-hit
and multi-hit models.

4.2.4 Linearized Multistage Model

This model which is an approximation derived from the multistage model is probably the most popular for dichotomous responses in cancer modeling. It is based on the assumption that several genetic changes are required to transform a normal cell to a cancer cell. The model is most widely used by the regulatory agencies and is given by

$$P(d) = 1 - e^{-\left(\beta_0 + \beta_1 d + \beta_2 d^2 + \cdots + \beta_k d^k\right)} \quad \beta_0, \beta_1, \ldots, \beta_k > 0 \quad (4.13)$$

where k is the number of stages. Clearly, the linearized multistage model is a generalization of the one-hit and the Weibull models. In practice, often a quadratic exponent suffices for an adequate fit to the data. We will discuss the multistage models of carcinogenesis in the next section of this chapter.

Example 4.2

Littlefield et al. (1980) describe an experiment conducted at the National Center for Toxicological Research (NCTR) in which BALB/c female mice were exposed to the drug 2-acetylaminofluorene (2-AAF) through diet. The number of animals in each dose group and the number dying with bladder carcinoma along with the associated proportion is given in Table 4.3. We will refer to this data set as the bladder carcinoma dose-response (BCDR) data set.

Note that as the dose increases, the number of animals in the non-zero dose groups decreases. This is because animals die of other causes as the exposure level rises. Additionally, even though a relatively large number of doses was used with a large number of animals in each dose group, it is still not possible to establish a specific dose-response model.

To fit a model, the simplest approach would be to make a transformation to linearize the function, if possible and apply linear regression. However, that

TABLE 4.3

Dose-Response Data for Proportion of Mice with Bladder Carcinoma Exposed to 2-Acetylaminofluorene (2-AAF) (Littlefield et al., 1980)

Dose	Number in Group	Number Affected	Proportion
0	360	3	0.00833
30	1432	15	0.01047
35	707	4	0.00566
45	347	4	0.01153
60	242	4	0.01653
75	248	4	0.01613
100	124	39	0.31452
150	119	88	0.73950

approach often does not lead to satisfactory estimates. For example, to fit the extended one-hit model, we can linearize as

$$\log(1 - P(d))^{-1} = \beta_0 + \beta_1 d.$$

Using a simple linear regression for the BCDR data set, we obtain the following estimates for β_0 and β_1

$$\hat{\beta}_0 = -0.3117, \quad \hat{\beta}_1 = .00865$$

with standard errors of respectively 0.1603 and 0.00212. But these estimates lack many of the desired statistical properties, especially since the correlations between the transformed probabilities are ignored. Improved estimates are derived by using the maximum likelihood method. However, this latter approach leads to nonlinear equations that need to be solved using a computer program. Using the binomial distribution, the likelihood is given by

$$L(\beta_0, \beta_1) = \prod_{i=0}^{g-1} \binom{n_i}{X_i} P_i^{X_i} (1 - P_i)^{n_i - X_i}$$

where g is the number of dose levels with $g_0 = 0$ being the background dose, P_i is the probability of response at dose i and X_i is the number of responses at that dose level. Now, p_i is replaced by an assumed model and the likelihood is maximized in order to derive the parameter estimates. To illustrate, once again assuming the extended one-hit model, the log-likelihood is given by

$$\log\{L(\beta_0, \beta_1)\} = \sum_{i=0}^{g-1} \log\binom{n_i}{X_i}$$

$$+ \sum_{i=0}^{g-1} X_i \log\left(1 - e^{-(\beta_0 + \beta_1 d_i)}\right)$$

$$- \sum_{i=1}^{g-1} (n_i - X_i)(\beta_0 + \beta_1 d_i)$$

Differentiating this function with respect to β_0 and β_1 yields

$$\frac{\partial}{\partial \beta_0} \log\{L(\beta_0, \beta_1)\} = \sum_{i=0}^{g-1} \frac{X_i e^{-(\beta_0 + \beta_1 d_i)}}{1 - e^{-(\beta_0 + \beta_1 d_i)}} - \sum_{i=0}^{g-1} (n_i - X_i)$$

$$\frac{\partial}{\partial \beta_1} \log\{L(\beta_0, \beta_1)\} = \sum_{i=0}^{g-1} \frac{d_i X_i e^{-(\beta_0 + \beta_1 d_i)}}{1 - e^{-(\beta_0 + \beta_1 d_i)}} - \sum_{i=0}^{g-1} d_i (n_i - X_i).$$

Upon setting the derivatives equal to 0, we get the following two likelihood equations

$$\frac{1}{N} \sum_{i=0}^{g-1} \frac{X_i}{1 - e^{-(\beta_0 + \beta_1 d_i)}} = 1 \tag{4.14}$$

$$\frac{1}{N} \sum_{i=0}^{g-1} \frac{d_i X_i}{1 - e^{-(\beta_0 + \beta_1 d_i)}} = \bar{d} \tag{4.15}$$

where $N = \sum_{i=0}^{g-1} n_i$ is the total number of animals in all dose groups and

$\bar{d} = \frac{1}{N} \sum_{i=0}^{g-1} n_i d_i$ is the weighted average dose. Equations (4.14) and (4.15) clearly cannot be solved analytically and a numerical algorithm such as the Fisher's method of scoring has to be used in order to find the roots. Here, we will not discuss the details for this method since there are a number of computer programs that can be applied to solve these equations. In fact, using the nlm() function in R, which is a nonlinear minimization function in R, we find that maximum likelihood estimates for β_0 and β_1 are respectively given as

$$\hat{\beta}_0 = -0.013986, \quad \widehat{\beta}_1 = .000843.$$

Similarly, we can use any of the other mechanistic dose-response models discussed above.

4.2.5 Multistage Models of Carcinogenesis

According to Whittemore and Keller (1978) perhaps the earliest attempt to model the process of carcinogenesis can be attributed to Iverson and Arley (1950) who postulated the one-stage theory of carcinogenesis. According to this one-stage theory, normal cells are transformed at a certain rate and become malignant. If we denote by $P_0(t)$ the probability that a cell is normal and at time t, and by $P_1(t)$ that it is malignant i.e. it is transformed at time t, then we can write (Whittemore and Keller, 1978)

$$\frac{dP_0}{dt} = -\lambda(t)P_0(t), \quad P_0(0) = 1 \tag{4.16}$$

$$\frac{d P_1(t)}{dt} = \lambda(t)P_0(t) \quad P_1(0) = 0 \tag{4.17}$$

where $\lambda(t)$ is the transition probability rate. From (4.16), we have

$$\frac{dP_0(t)}{P_0(t)} = -\lambda(t)dt$$

Integrating both sides and imposing the initial condition, we obtain

$$P_0(t) = \exp\left(-\int_0^t \lambda(s)\,ds\right)$$

as the solution of (4.16). Similarly noting that $P_0(t) + P_1(t) = 1$, we find the solution of (4.17) as

$$P_1(t) = 1 - \exp\left(-\int_0^t \lambda(s)\,ds\right) \tag{4.18}$$

Let T_1 and W be the transformation time and growth time of the tumor with distributions $g_1(t)$ and $g_W(t)$ respectively. Then assuming T_1 and W are independent, the distribution of the total time to detection of tumor $T = T_1 + W$ is the convolution of $h_1(t)$ and $h_W(t)$ given by

$$g(t) = \int_0^t g_1(u)g_W(t-u)\,du \tag{4.19}$$

Now, from (4.18) we can find $g_1(t)$ as the transformation rate given by

$$g_1(t) = \frac{dP_1(t)}{dt} = \lambda(t)\exp\left(-\int_0^t \lambda(s)\,ds\right) \tag{4.20}$$

Substituting in (4.19), we get

$$g(t) = \int_0^t \lambda(u)\exp\left(-\int_0^u \lambda(s)\,ds\right)g_W(t-u)\,du \tag{4.21}$$

Once the transition rate $\lambda(t)$ and the tumor growth distribution $g_W(t)$ are specified, then $g(t)$ can be determined from (4.21). Iverson and Arley (1950) assume that $\lambda(t)$ is a linear function of the dose and that the growth of the tumor is governed by a pure birth process with a constant birth rate. Specifically, if it can be assumed that λ is constant, then (4.21) reduces to

$$g(t) = \int_0^t \lambda e^{\lambda u} g_W(t-u)\,du \tag{4.22}$$

The one-stage theory of Iverson and Arley was soon overshadowed by the fact that many investigators observed that the incidence rate was proportional to a certain power (about fifth or sixth) of exposure duration. This led to the belief that a single cell goes through a finite number of stages to become malignant and generate a tumor. The pioneers of this theory were Muller (1951) and Nordling (1953).

The multistage theory is based on the assumption that cancer is the result of a sequence of irreversible genetic transformations. A normal cell can become malignant only after gong through a series of mutational changes induced by exposure. This theory was utilized by Armitage and Doll (1954) to derive a mathematical description of the cancer process and derive the Armitage–Doll multistage model, which is still a widely accepted model for cancer. We outline the derivation of the Armitage–Doll model. For more details, we refer to Meza (2010).

Suppose that a normal cell would have to go through n transformations (stages) to become malignant (see Figure 4.2).

Let $P_j(t); j = 0,\ldots,n$ be the probability that the cell is in stage j at time t. Then the process of cell malignancy can be thought of as a pure birth process with an absorbing barrier at the nth stage. By using the Chapman-Kolmogorov equations (see for example Ross, S., 2019), we can express the state probabilities $P_j(t)$ as the following set of equations:

$$P_0(t+h) = (1 - h\lambda_0)P_0(t) + o(h)$$

$$P_j(t+h) = (1 - h\lambda_j)P_j(t) + h\lambda_{j-1} P_{j-1}(t) + o(h) \quad j = 1,\ldots,n$$

where λ_k is defined to be 0. By rearranging the above equations, dividing by h, and taking the limit as $h \to 0$, we get

$$\frac{d P_0(t)}{dt} = -\lambda_0 P_0(t)$$

$$\frac{d P_j(t)}{dt} = -\lambda_j P_j(t) + \lambda_{j-1} P_{j-1}(t) \quad j = 1,\ldots,n-1 \qquad (4.23)$$

$$\frac{dP_n(t)}{dt} = \lambda_{n-1} P_{n-1}(t)$$

Note that the set of $(n + 1)$ differential equations represented in (4.23) are the natural extension of (4.16) and (4.17) with constant transition rates $\lambda_0, \lambda_1, \ldots,$

FIGURE 4.2
Multistage model.

λ_n. In the more general case, the rates may be time dependent, but for simplicity, here we assume they are constants. The solution of the above system of equations may be derived as

$$P_0(t) \propto e^{-\lambda_0 t}$$

$$P_j(t) = e^{-\lambda_j t} \int_0^t \lambda_{j-1} P_{j-1}(u) e^{\lambda_j u}\, du \quad j = 1, \ldots, n$$

Which by repeated integration leads to

$$P_j(t) \propto \frac{\lambda_0 \lambda_1 \ldots \lambda_{j-1}}{j!} t^j \quad j = 1, \ldots, n$$

Now, if we assume that there is a total of N susceptible cells independently going through full malignancy transformation at times $T_1, T_2, \ldots, T_N$ respectively and if $T = \min(T_1, T_2, \ldots, T_N)$ is the time to cancer development, the survival function for T is given by

$$S(t) = P(T > t) = (1 - P_n(t))^N \tag{4.24}$$

and the overall hazard for malignancy is given by

$$h(t) = -\frac{S'(t)}{S(t)} = \frac{N P_k'(t)}{1 - P_k(t)} \tag{4.25}$$

By writing $P_n(t)$ as the cumulative distribution function of a random variable which is the sum of k independent exponential variables with rates $\lambda_0, \ldots, \lambda_{n-1}$, Moolgavkar (1991) shows that the problem of computing $P_k'(t)$ reduces to the convolution of k independent exponentially distributed random variables. If we assume further that the malignancy at the cell level occurs with negligible probability, i.e. $P_k(t) \approx 0$, then by using Taylor expansion, we can show that the hazard function $h(t)$ can be expressed as approximately

$$h(t) = \frac{N \lambda_0 \lambda_1 \ldots \lambda_{n-1}}{(n-1)!} t^{n-1} \left\{ 1 - \bar{\lambda} t + f(\lambda, t) \right\}$$

where $\bar{\lambda}$ is the mean of the transition rates, i.e. $\bar{\lambda} = \frac{1}{k} \sum_i^{k-1} \lambda_i$ and $f(\lambda, t)$ is a function that involves second- and higher-order terms in the product of transition rates. Further, by keeping only the first non-zero term of the Taylor series, we obtain

$$h(t) \approx \frac{N \lambda_0 \lambda_1 \ldots \lambda_{n-1}}{(n-1)!} t^{n-1} \tag{4.26}$$

which is the Armitage–Doll approximation of the hazard function. This approximation indicates that the age-specific hazard rate is proportional to the power of age. This result is consistent with the epidemiological observed rates in many different types of cancer. In fact Armitage–Doll (1985) states that if the number of distinct stages is between 2 and 6, the agreement with the epidemiological results is quite good.

Denoting the cumulative distribution of the time to tumor development T by $F(t)$, i.e. the probability of having cancer by time t, then we have

$$h(t) = \frac{f(t)}{1 - F(t)} = -\frac{d}{dt} \ln\left(1 - F(t)\right)$$

and so

$$\int_0^t h(u)\,du = -\ln\left(1 - F(u)\right)\Big|_{u=0}^{u=t}$$

$$= -\ln\left(1 - F(t)\right)$$

Thus

$$F(t) = 1 - e^{-\int_0^t h(u)\,du} \tag{4.27}$$

Substituting from (4.26), we get

$$F(t) = 1 - e^{-\Lambda t^n} \quad t \geq 0 \tag{4.28}$$

with $\Lambda = \dfrac{N \lambda_0 \lambda_1 \ldots \lambda_{n-1}}{n!}$.

This is the cumulative distribution function of the Weibull model. Thus the hazard function given in (4.26) corresponds to a Weibull model. The significance of the Armitage–Doll model is that it demonstrates that the hazard (at least approximately) behaves as the power of age, which is observed to be consistent with many types of cancer. On the other hand, a major limitation of the model is that it does not allow for cell death implying that any susceptible cell will eventually become a cancer cell. However, because of its simplicity and wide applications, the model is one of the most commonly used tools for cancer modeling. Note also that if we further assume that the each transition rate, λ_i is a linear function of dose, i.e.

$$\lambda_i(d) = \alpha_i + \beta_i d \quad i = 0, 1, \ldots, n-1$$

then from (4.8) the probability distribution of time to tumor, that is, the dose-response model associated with the Armitage–Doll model, can be written as

$$P(t,d) = 1 - e^{-N(\alpha_0 + \beta_0 d)\ldots(\alpha_{n-1} + \beta_{n-1}d)t^n}$$

which can be simplified as

$$P(t,d) = 1 - \exp\left[-Q(d)t^n\right]$$

where $Q(d)$ is a polynomial of degree n in d with coefficients determined by the product of the n linear functions $\alpha_i + \beta_i d$. In toxicological bioassay experiments, the animal exposure is through lifetime, and therefore t is a constant and the dose-response model reduces to

$$P(d) = 1 - \exp\left[-\left(q_0 + q_1 d + \cdots + q_n d^n\right)\right] \quad q_0, q_1, \ldots, q_n \geq 0 \qquad (4.29)$$

This model has been extensively used in toxicological dose-response model and risk assessment. In practice, often a liner function or a polynomial of degree 2 suffices and it is shown that not much is gained in using polynomials of higher degree. Using the US Environmental Protection Agency (EPA) Integrated Risk Information System (IRIS) database, Nicheva et al. (2007) showed that in animal carcinogenicity experiments the addition of the second-order term improves the fit of the model in only 20% of cases and addition of higher-order terms does not contribute to the fit of the model at all.

One of the mathematical properties of the dose-response model in (4.29) in risk assessment is that if the relative risk defined in Chapter 2 as

$$R(d) = \frac{P(d) - P(0)}{1 - P(0)}$$

is used as the measure of risk, then by substituting from (4.29) we have

$$R(d) = 1 - \exp\left[-\left(q_1 d + \cdots + q_n d^n\right)\right] \qquad (4.30)$$

By expanding the exponential term, we get

$$R(d) = 1 - \left[1 - q_1 d + O(d^2)\right] \qquad (4.31)$$

Now, since humans are typically exposed to much lower levels of carcinogens than the dose levels in animal bioassay experiments, to determine the risk at the low human dosage levels, one needs to extrapolate from the dose-response model below the lowest experimental non-zero dosage level. Equation (4.21) shows that, in practice, to estimate the cancer risk for low-dose extrapolation, i.e. as d approaches 0, ignoring the terms involving the quadratic and higher orders of d, the practical implication is that one needs to consider only the slope of the line connecting the lowest animal exposure level to the origin. That is, for small values of d we have

$$R(d) \approx q_1 d \tag{4.32}$$

This is called the linearized multistage model which has been used frequently for low-dose extrapolation (see Crump (2018) and references therein). Equation (4.15) further means that the marginal increase in cancer risk for small doses is approximately equivalent to the slope of the risk function at that dose level. This clearly implies that the rise in risk for small increase in exposure level at a given dose is determined by the first derivative of the risk function evaluated at the given dose level.

4.2.5.1 Solution of the Armitage–Doll Model

By solving the system of differential equations (4.3), the exact solution for the survival and hazard functions of the Armitage–Doll model can be derived. As noted by Meza (2010), the solution depends on eigenvalues of the system, which in this situation are exactly the transition rates of the model. If all the transition rates are different and no two of the transition rates are equal, then (Meza, 2010) survival and hazard functions are respectively given by

$$S(t) = \left[\frac{\lambda_0 \ldots \lambda_{n-1}}{L^*} \sum_{k=0}^{n-1} \frac{e^{-\lambda_k t}}{\lambda_k} L_k \right]^N$$

$$h(t) = \frac{N \sum_{k=0}^{n-1} e^{-\lambda_k t} L_k}{\sum_{k=0}^{n-1} \frac{e^{-\lambda_k t}}{\lambda_k} L_k},$$

where

$$L^* = \prod_{i<j} (\lambda_i - \lambda_j),$$

$$L_k = \prod_{\substack{i,j \neq k \\ i<j}} (\lambda_i - \lambda_j).$$

In the special case when all transition rates are equal, the survival and hazard functions are given by

$$S(t) = e^{-\lambda t} \sum_{k=0}^{n-1} \frac{(\lambda t)^k}{k!}$$

$$h(t) = \frac{N t^{n-1} \lambda^n}{\sum_{k=0}^{n-1} \frac{(n-1)!}{k!} (\lambda t)^k}.$$

4.2.6 Two-Stage Clonal Expansion Model

The Armitage–Doll multistage model is probably the most widely used model in animal carcinogenicity studies. The mathematical simplicity and biological plausibility of the model has attracted the attention of many toxicologists. However, the model is based on certain assumptions that, if violated, could make the model completely invalid. For example, Moolgavkar (1978) points out that if the transition rates are not sufficiently small, the Armitage–Doll model is inappropriate. The two-stage clonal expansion (TSCE) model which was first developed by Moolgavkar and Venzon(1979) and Mookgavkar and Knudson(1981), is based on specifics of cell kinetics such as cell growth and death or differentiation and provides a plausible alternative to the multistage model. The original model has undergone some extensive development to make it amenable for studying various human cancers and for applications in tumor promotion studies. The TSCE is based on the assumption that the transition of a normal cell to a malignant cell is the result of two irreversible events. A normal cell is first changed into an initiated or intermediate cell. This is a rare event and it is a process in which normal cells are transformed so that they are able to form tumors and it is considered to be the first phase in tumor development. This process is usually modeled as a non-homogeneous time-dependent Poisson process. In the second or the promotion stage, the initiated cells undergo clonal expansion which means that the cells either are divided into two initiated cells, die or are differentiated, or divide into initiated and malignant cells. The conversion of the initiated cell to a malignant cell is also called progression and can involve one or more mutations. This process is modeled as a stochastic birth-death process. Figure 4.3 depicts the process of TSCE.

The TSCE model has been considered by many authors. Not only have several investigators discussed the biological plausibility application of the model in different types of cancer (see e.g., Kruse et al., 2002), the statistical properties and various extensions have also been the subject of numerous research papers (e.g., Brouwer et al., 2017). Here, we outline the general mathematical derivation of the TSCE model, and for more details we refer to Moolgavkar and Luebeck (1990).

Let $X(t)$ be the number of normal susceptible cells at time t. In its most general form, the TSCE model assumes that $X(t)$ is a stochastic process. But, for simplicity, here we consider only the situation where the normal stem cells follow a deterministic process. As mentioned before, the process of initiation of normal cells into intermediate cells is modeled as a non-homogeneous Poisson process. Let $\nu(t)$ be the first mutation rate per cell at time t. Then, $\nu(t)$

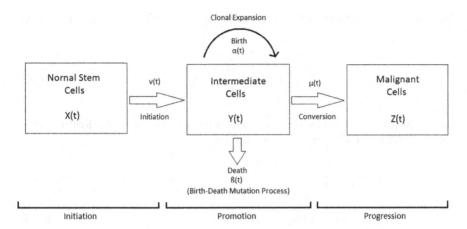

FIGURE 4.3
Two-stage clonal expansion model (Meza, 2010).

$X(t)$ is the intensity of the Poisson process. Now, according to the TSCE model the initiated cells divide into two normal cells, die, or divide into a normal and a malignant cell. Let the rates at which the three events occur be respectively $\alpha(t)$, $\beta(t)$, and $\mu(t)$. Let $Y(t)$ and $Z(t)$, respectively, denote the number of intermediate and malignant cells at time t. Let

$$P_{j,k}(t) = \text{Prob}\big[Y(t) = j, Z(t) = k \mid Y(0) = Z(0) = 0\big]. \qquad (4.33)$$

Then, the vector $\{Y(t), Z(t)\}$ can be regarded as a two-dimensional Markov process and therefore it satisfies the Chapman-Kolmogorov equation. Thus we have

$$P_{j,k}(t+h) = \big[v(t)X(t)P_{j-1,k}(t) + (j-1)\alpha(t)P_{j-1,k}(t)\big]h$$

$$+ \big[(j+1)\beta(t)P_{j+1,k}(t) + j\mu(t)P_{j,k-1}(t)\big]h$$

$$+ \big[1 - jh(\alpha(t) - \beta(t) - \mu(t)) - hX(t)v(t)\big]P_{j,k}(t) + o(h)$$

Subtracting $P_{j,k}(t)$ on both sides of the above equation, dividing by h, and taking the limit as h approaches 0, we get the Kolmogorov differential equation (see Moolgavkar et al., 1988) as

$$\frac{d}{dt}P_{j,k}(t) = \big[v(t)X(t) + (j-1)\alpha(t)\big]P_{j-1,k}(t)$$

$$+ (j+1)\beta(t)P_{j+1,k}(t) + j\mu(t)P_{j,k-1}(t)$$

$$- j\big(\alpha(t) - \beta(t) - \mu(t)\big) + v(t)X(t)P_{j,k}(t)$$

Now, let $\Psi(y, z; t)$ denote the joint probability generating function of $Y(t)$ and $Z(t)$, that is

$$\Psi(y,z;t) = \sum_{j=0}^{\infty}\sum_{k=0}^{\infty} P_{j,k}(t)y^j z^k \tag{4.34}$$

then multiplying (4.34) by $y^j z^k$ and summing over j and k, the Kolmogorov forward differential equation is resulted as

$$\Psi'(y,z;t) = \frac{\partial \Psi(y,z;t)}{\partial t}$$

$$= (y-1)v(t)X(t)\Psi(y,z;t) \tag{4.35}$$

$$+ \left\{\mu(t)yz + \alpha(t)y^2 + \beta(t) - [\alpha(t) + \beta(t) + \mu(t)]y\right\}\frac{\partial \Psi}{\partial y}$$

Assuming that all cells are initially normal, the initial condition for the above equation is $\Psi(y, z; 0) = 1$. Note also that $\Psi(1, 0; t) = P(Z(t) = 0)$ is the survival function $S(t)$ for time T to the first malignant cell. The corresponding hazard function is given by

$$h(t) = -\frac{S'(t)}{S(t)} = \frac{\Psi'(1,0;t)}{\Psi(1,0;t)}.$$

Therefore, we see that in order to derive expressions for the survival and hazard functions, we need to evaluate $\Psi(1, 0; t)$ and its derivative. Note that from (4.35) above, we have

$$\Psi'(y,z;t) = -\mu(t)\frac{\partial \Psi(1,0;t)}{\partial y}$$

and therefore

$$h(t) = \mu(t)E[Y(t)Z(t) = 0]$$

since

$$E[Y(t) \mid Z(t) = 0] = \frac{1}{\Psi(1,0;t)}\frac{\partial \Psi(1,0;t)}{\partial y}.$$

Now, as pointed out before, since the occurrence of cancer is a rare event, i.e. $P[Z(t) = 0] \approx 1$, we can replace the above conditional expectation with the unconditional expectation. Substituting in (4.35) and solving, we finally find the following approximation for the hazard function

$$h(t) \approx \mu(t)E[Y(t)] = \mu(t)\int_0^t v(s)X(s)\exp\left[\int_s^t (\alpha(u) - \beta(u))du\right]ds \qquad (4.36)$$

Dewanji (2008) notes that the above approximation besides being inapplicable in animal carcinogenicity studies, also lacks flexibility as it depends on two parameters, namely the net cell proliferation rate $(\alpha - \beta)$ and the product $\mu.vX$. In other words, the above solution gives us no information about the role of parameters α and β alone. Also, in experimental data, often the probability of malignancy is not small and therefore approximating the conditional expectation with the unconditional one may lead to inaccuracy. The hazard calculated from the above approximation and the exact solution could be very different in carcinogenesis experiments.

4.2.6.1 *Exact Hazard for Piecewise Constant Rate Parameters*

In its most general form, the model parameters are age-dependent and in order to evaluate the carcinogenic effects with age, we need to have solutions for the hazard and survival functions. However, closed form solutions for these functions are not available for age-dependent parameters. But, it turns out that the computation of the hazard and survival functions is possible when the rate parameters are piecewise constant. In this case, the model leads to a recursive procedure that results in closed form expressions for the hazard and survival functions of the TSCE model. The complete derivation of the solution is complex and involves solving a certain Ricatti differential equation. Here, we give an outline and refer the reader to Heidenreich et al. (1997). In practice, constant rate parameters most frequently occur in epidemiologic data and indeed in animal bioassay experiments.

Suppose therefore that the parameters α, β, and v are constant within the time interval $[t_{i-1}, t_i]$, i.e.

$$\begin{cases} \alpha(t) = \alpha_i \\ \beta(t) = \beta_i \qquad t_{i-1} \le t \le t_i \quad i = 1, \ldots, n \\ v(t) = v_i \end{cases}$$

where we note that the time interval $[0,t]$ is broken down to n subintervals $[t_{i-1}, t_i]$ for $i = 1, \ldots, n$. Now, the derivation of the exact solution of the Kolmogorov equation (4.15) requires the associated characteristic equations that are given by

$$\frac{dy}{ds} = -\left\{ \left[\mu(s)z + \alpha(s)y - \alpha(s) - \beta(s) - \mu(s) \right] y + \beta(s) \right\}$$

$$\frac{dz}{ds} = 0, \ \frac{dt}{ds} = 1, \ \frac{d\Psi}{ds} = (y-1)v(s)X(s)\Psi.$$

In order to derive the hazard, as was shown earlier, we need to only solve for $\Psi(1, 0, t)$. This means that we only need to find $\Psi(y, z; t)$ along the characteristic through $(y(0), 0, 0)$ with $y(t) = 1$. Now, the Kolmogorov ordinary differential equation can be solved along characteristics, which gives the following solution:

$$\Psi(y(t), z; t) = \Psi_0 \exp\left\{\int_0^t [y(s,t)-1]v(s)X(s)ds\right\}$$

for which the initial condition is $\Psi_0 = \Psi(y(0), z(0); 0) = 1$. The equation for $y(s, t)$ is a Ricatti differential equation which can be solved analytically when the model parameters are piecewise constant. In this case the solution to the characteristic equation is (Heidenreich et al., 1997; Meza, 2010)

$$\alpha_i(y(s,t)-1) = \frac{\partial \log f_i(s,t)}{\partial s}$$

where

$$f_i(s,t) = (\tilde{y}_i - p_i)\exp\left[q_i(s-t_i)\right] + (q_i - \tilde{y}_i)\exp\left[p_i(s-t_i)\right]$$

with p_i and q_i being the lower and upper roots of the quadratic equation

$$U^2 + (\alpha_i - \beta_i - \mu_i)U - \alpha_i\mu_i = 0$$

and

$$\tilde{y}_{i-1} = \frac{\alpha_{i-1}}{\alpha_i} \frac{\partial \log f_i(s,t)}{\partial s}\bigg|_{s=t_{i-1}}.$$

From the expressions, the equations for the hazard and survival functions when the rate parameters are age-dependent can be derived as

$$S(t) = \exp\left[\sum_{i=1}^{n} \frac{v_i}{\alpha_i} \log \frac{q_i - p_i}{f_i(t_{i-1}, t)}\right]$$

$$h(t) = \sum_{i=1}^{n} \frac{v_i}{\alpha_i} \frac{\partial \log\left[f_i(t_{i-1}, t)\right]}{\partial t}.$$

4.3 Discussion

In dose-response modeling of cancer from animal experiments, there has been a long controversy as to whether or not a threshold dose exists below which no adverse effect can occur. Some toxicologists believe that threshold doses appear to exist for direct-acting carcinogens and mutagens and even if a threshold dose does exist for a chemical, it would be hard to estimate it from animal bioassay experiments. Recall that in order to detect a potential adverse effect to chemical exposures with a limited number of animals, bioassay experiments are designed so that exposure occurs at dosage levels that are considerably higher than the human levels, and to predict the health effects in low doses, it is necessary to extrapolate from the lowest exposure level to the background control level. It is to be noted that there are actually two types of thresholds and it is necessary to distinguish between individual and population thresholds. Clearly there is variability in tolerance to chemicals for animals in the population. Thus, we can say that a population threshold applies to a large number of animals. That is to say, that a threshold dosage level for the population is a dosage such that no animal in the population would not exhibit adverse effect, thus being the minimum of the individual animal threshold levels. Although animal bioassay experiments may lead to the conclusion that there are individual thresholds since within the same dose group, some animals develop tumor while others do not, there are several arguments that this variability may not be due to the existence of individual thresholds and there may be several other reasons. Even after fitting a dose-response model, it is observed that as dosage is decreased, the prevalence of an effect reduces to zero. Gaylor (1985) presents several arguments as to why it is difficult to establish the existence of a threshold through animal carcinogenicity experiments. In that paper, the following steps are given to determine an upper confidence limit for risk by extrapolating below the lowest experimental dosage level:

1. Fit a mathematical dose-response model that gives zero excess tumor at zero dose. Check the fit and adequacy of the model.
2. Determine the upper confidence limits on risk, i.e. the excess tumor rate above the control animals in the experimental doses.
3. Connect a straight line from the origin to the point on the upper confidence level at the lowest experimental dosage.
4. Find the upper confidence limit on risk at a desired low-dose level. Alternatively, determine dosage corresponding to a low nominal risk level using the straight line.

This method of extrapolation was first proposed by Gaylor and Kodell (1980). The advantage of the method is that the use of an upper confidence limit somewhat reduces the uncertainty in model fitting as there may be several models that can adequately fit the data.

Exercises

1. In the one-stage theory of carcinogen discussed in Section 4.2 of this chapter, the distribution of the time to a detectable tumor is given by (4.22). Iverson and Arley (1950) assumed that the tumor growth is governed by a pure birth process with rate γ Then the probability $Q(n, t)$ that a clone consists of m cells after the a time of t is given by

    $$\frac{d}{dt}Q(n,t) = -\gamma n\, Q(n,t) + \gamma(n-1)Q(n-1,t) \qquad (4.37)$$

    The solution of (4.37) is given by (Whittemore and Keller, 1978)

    $$Q(n,t) = e^{-\gamma t}\left(1 - e^{-\gamma t}\right)^{n-1} \qquad (4.38)$$

    Verify that (4.38) satisfies (4.37) and use it along with (4.22) to find the distribution of time to tumor.

2. Based on the Armitage–Doll model, show that the mean and variance of time to tumor development are respectively given by

    $$\mu = \frac{1}{\Lambda^{1/n}}\Gamma\left(1 + \frac{1}{n}\right)$$

    and

    $$\sigma^2 = \frac{1}{\Lambda^{2/n}}\left\{\Gamma\left(1 + \frac{2}{n}\right) - \left[\Gamma\left(1 + \frac{1}{n}\right)\right]^2\right\}.$$

3. Consider the differential equation

    $$U'(t) = \frac{dU}{dt} = \alpha U^2(t)(\alpha + \beta + \mu)U(t) + \beta \qquad (4.39)$$

    with the initial condition $U(0) = 1$.
    (i)   Show that

    $$\alpha dt = -\frac{1}{q-p}\left(\frac{dU}{q-U} - \frac{dU}{dp}\right)$$

where p and q are, respectively, the lower and upper roots of the quadratic equation (4.39).

(ii) Integrating the expression in (i), show that

$$U(t) = p + \frac{q-p}{1 + \dfrac{q-1}{1-p}\exp(\alpha(q-p)t)} \qquad (4.40)$$

4. In a generalization of the TSCE model, one can think of the process as consisting of k stages with a fixed number of stem cells N. Suppose each cell mutates at rate μ_0 and becomes a type 1 cell. Thus type 1 cells are produced as a Poisson process with rate $\gamma = N\,\mu_0$. Similarly, the pre-initiated cells of types $i = 1, \ldots, k-2$ mutate at rate $\mu_{i,\,k}$ and become type $i+1$. At the final stage, cells of type $k-1$ are initiated cells that divide at rate α, die at rate β, and mutate and become malignant at rate $\mu_{k-1,\,k}$. Then if T_k is the time to the first malignant cell, then we can show (Lian and Durrette, 2018) that the survival function is given by

$$S_k(t) = \exp\left\{ -\gamma \int_0^t \left[1 - W_{k-1,k}(t-s) \right] ds \right\}$$

where $W_{i,\,k}(t)$ is a function recursively given by

$$W_{i,k}(t) = \exp\left\{ -\mu_{k-i,k} \int_0^t \left[1 - W_{i-1,k}(t-s) \right] ds \right\}$$

with $W_{1,\,k}(t)$ given by the right hand side of (4.40). Show that

(i) The hazard function is given by

$$h_k(t) = \gamma\left(1 - W_{k-1,k}(t) \right).$$

(ii) Specifically for $k = 2, 3, 4$, show that

$$h_2(t) = \frac{\gamma}{\alpha}\,\frac{-PQ\left(e^{-Pt} - e^{-Qt}\right)}{Qe^{-Pt} - Pe^{-Qt}}$$

$$h_3(t) = \gamma\left\{ 1 - \left[\frac{Q-P}{Qe^{-Pt} - Pe^{-Qt}} \right]^{\frac{\mu_1}{\alpha}} \right\}$$

$$h_4(t) = \gamma \left\{ 1 - \exp \left[-\mu_1 \int_0^t \left(1 - \left(\frac{Q - P}{Qe^{-P(t-s)} - Pe^{-Q(t-s)}} \right)^{\frac{\mu_2}{\alpha}} \right) \right] \right\}$$

where in the above expressions, $P = \alpha(p - 1)$ and $Q = \alpha(q - 1)$.

5. In this example, we use the Gaylor–Kodell (1980) procedure for linear extrapolation to compare two dose-response models. Littlefield et al. (1980) describe a large-scale experiment conducted at the National Center for Toxicological Research (NCTR) in which 2-acetylamino-flourence (2-AAF) was fed to BALB/c female mice in order to assess the urinary bladder carcinoma dose-response (BCDR). Table 4.4 gives the experimental dose levels and the number of animals per group along with the number of animals dying from exposure to the chemical.

   a. Fit the Armitage–Doll multistage model with a quadratic exponent

   $$P(d) = 1 - \exp \left[-\left(\beta_0 + \beta_1 d + \beta_2 d^2 \right) \right] \tag{4.41}$$

   to the data and assess the goodness of fit.

   b. Find the upper 97.5% confidence limit on the excess bladder carcinoma rate at 30 ppm.

   c. Find the equation of the line of interpolation from the upper confidence limit at the dosage level of 30 ppm to the origin.

   d. Find an estimate of the dosage level that gives at least 97.5% confidence that the excess bladder carcinoma rate does not exceed one in million.

   e. Razzaghi (2002) introduces the mixture modes as a flexible alternative to dose-response modeling. Fit a mixture of two multistage models with six parameters.

TABLE 4.4

Dose-Response Data for BALB/c Mice Exposed to 2-Acetylaminofluorene (2-AAF)

Dosage (ppm)	Number of Animals	Number Affected
0	360	3
309	1432	15
35	707	4
45	347	4
60	242	4
75	248	20
100	124	39
150	119	88

$$P(d) = \theta \left\{ 1 - \exp\left[-\left(\gamma_0 + \gamma_{10}\, d + \gamma_{02}\, d^2 \right) \right] \right\}$$

$$+ (1 - \theta)\left\{ 1 - \exp\left[-\left(\gamma_0 + \gamma_{11}\, d + \gamma_{22}\, d^2 \right) \right] \right\}$$

(4.42)

to the data and assess the goodness of fit.

f. Repeat parts b–d for Model (4.42).

g. Compare and comment.

It is a transformational process during which normal cells go through a series of genetic and epigenetic changes and are transformed into malignant and cancer cells. Although the specific stages of transformation are not completely known, there is evidence to suggest that the general framework of change can be formulated as initiation-promotion-malignancies. Initiation refers to the stage where the normal cells are altered and become prone to clonal expansion. Promotion is the stage where the initiated cells transform into premalignant cells. And finally malignancy is the stage of the growth of malignant cells into tumors and the onset of cancer.

References

Armitage, P. (1985). Multistage models of carcinogenesis. *Environmental Health Perspectives*, 63, 195–201.

Armitage, P. and Doll, R. (1954). The age distribution of cancer and a multistage theory of carcinogenesis. *British Journal of Cancer*, 8, 1–12.

Brouwer, A. F., Meza, R., and Eisenberg, M. C. (2017). A systematic approach to determining the identifiability of multistage carcinogenesis models. *Risk Analysis*, 37, 1375–87.

Crump, K. (2018). Cancer risk assessment and the biostatistical revolution of the 1970s – A reflection. *Dose Response*, 16. doi:10.1177/1559325818806402.

Dewanji, A. (2008). Models for carcinogenesis. In *Statistical advances in the biomedical sciences*, A. Biswas, S. Datta, and Fine, J. P. eds. Wiley and Sons, New York, 547–68.

Fjeld, R. A., Eisenberg, N. A., and Compton, K. (2006). *Quantitative environmental risk analysis for human health*. Wiley and Sons, Hoboken, 258.

Gaylor, D. W. (1985). The question of existence of thresholds: Extrapolation from high to low dose. In *Advances in modern environmental toxicology, Volume XII, Mechanisms and toxicity of chemical carcinogens and mutagens*, W. G. Flamm and R .J. Lorentzen eds. Princeton Scientific, Princeton, 249–60.

Gaylor, D. W. and Kodell, R. L. (1980). Linear interpolation algorithm for low dose risk assessment of toxic substances. *Journal of Environmental and Pathological Toxicology*, 4, 306–12.

Heidenreich, W. F., Luebeck, E. G., and Moolgavkar, S. H. (1997). Some properties of the hazard function of the two-mutation clonal expansion model. *Risk Analysis*, 17, 391–99.

Iverson, S. and Arleys, N. (1950). On the mechanism of experimental carcinogenesis. *Acta Pathologica, Microbiologica, et Immunologica Scandinavica*, 27, 773–803.

Kruse, C., Jaedicke, A., Beaudouin, J., Bohl, F., Ferring, D., Guttier, T, Ellenberg, J., and Jansen, R. P. (2002). Ribonucleoprotein-dependent localization of the yeast class V myosin Myo4p. *Journal of Cell Biology*, 159, 971–82.

Lian, T. and Durrette, R. (2018). A new look at the multi-stage models of cancer incidence. https://doi.org/10.1101/243972.

Littlefield, N. A., Farmer, J. H., Gaylor, D. W., and Sheldon, W. G. (1980). Effects of dose and time in a long-term low dose carcinogenic study. *Journal of Environmental and Pathological Toxicology*, 3(Spec No), 17–34.

Meza, R. (2010). Stochastic models of carcinogenesis. Semanticscholor.org.

Moolgavkar, S. H. (1978). The multistage theory of carcinogenesis and the age distribution of cancer in man. *Journal of National Cancer Institute*, 61, 49–52.

Moolgavkar, S. H. (1991). Carcinogenesis models: An overview. *Basic Life Sciences*, 58, 387–396; with discussion.

Moolgavkar, S. H., Dewanji, A., and Venzon, D. J. (1988). A stochastic two-stage model for cancer risk assessment. I. The hazard function and the probability of tumor. *Risk Analysis*, 8, 383–92.

Moolgavkar, S. H. and Knudson, A. G. (1981). Mutation and cancer: A model for human carcinogenesis. *Journal of the National Cancer Institute*, 66, 1037–52.

Moolgavkar, S. H. and Luebeck, E. G. (1990). Two-event model for carcinogenesis: Biological, mathematical and statistical considerations. *Risk Analysis*, 10, 321–41.

Moolgavkar, S. H. and Venzon, D. J. (1979). Two-event model for carcinogenesis: Incidence curves for childhood and adult tumors. *Mathematical Biosciences*, 47, 55–77.

Muller, H. J. (1951). Radiation damage to the genetic material. In *Science in progress: Seventh series*, G. A. Baitsell ed. Yale University Press, New Haven, 93–165.

Nicheva, D., Piegorsch, W. W., and West, R. W. (2007). On use of the multistage dose-response model for assessing laboratory animal carcinogenicity. *Regulatory Toxicology and Pharmacology*, 48, 135–47.

Nordling, C. O. (1953). A new theory on cancer-inducing mechanism. *British Journal of Cancer*, 7, 68–72.

Quast, J. F., Wade, C. E., Humiston, C. G., Carreon, R. M., Mermann, E. A., Park, C. N., Schwetz, B. A. (1980). *A two year toxicity and oncogenicity study with acrylonitrile incorporated in the drinking water of rats*. Dow Chemical Co, Toxicology Research Laboratory, Midland, MI.

Razzaghi, M. (2002). The use of distribution mixtures for dose-response modeling in toxicological experiments. *Environmetrics*, 13, 1–11.

Ross, S. (2019). *Introduction to probability models*, 12th edition. Academic Press.

Whittemore, A. and Keller, J. B. (1978). Quantitative theories of carcinogenesis. *SIAM Review*, 20, 1–30.

5

Models in Reproductive and Developmental Toxicology

5.1 Introduction

The industrialization of the modern world has witnessed a steady growth of the usage of chemical agents in many different settings. Humans are routinely exposed to various chemicals through drugs, food additives, pollution, industrial waste, and other sources. At the same time the growth of scientific knowledge has produced vivid evidence that prenatal exposure to toxic agents can lead to adverse developmental effects. The science of toxicology has a long history, and the study of effects of toxic chemicals on the human body goes back to thousands of years. Specifically, several chemicals have been identified as toxicants causing undesirable effects on humans. However, the study of the developmental effects of toxic agents and how maternal exposure to chemicals during pregnancy can affect the developing fetus is fairly new and has attracted the attention of many scientists in the last 50 years. This is mainly due to the fact that for a long time toxicologists and other scientists believed that the placenta provided total protection for the embryo. It is only recently that scientists have realized that chemical agents can cross the placenta and severely harm the embryo. That is mainly the reason that in the last several decades thousands of experiments have been performed on laboratory animals to try and understand the way that chemicals can enter the placenta and the mechanism through which they can affect the developing fetus. These experiments have led to numerous attempts to identify teratogenic chemicals, and the results of these studies are used in risk assessment and estimation of risks to humans as the result of exposure to developmental toxicants. Today, there is little doubt that a woman's exposure to environmental toxins during pregnancy can severely harm the mother and the baby.

Reproductive and developmental toxicology is the branch of science that is the study of adverse developmental effects of drugs, chemicals, diet, and environmental agents (Mattison et al., 2003). Developmental toxicologists deal with problems of fertility, pregnancy, and causes, manifestation, mechanisms of action, and prevention of developmental abnormalities. The abnormalities, produced by developmental toxicants, may be structural or functional in nature. A developmental toxicant is defined as a chemical, drug, virus, bacteria, or any other agent that can alter the morphology or the subsequent function of a developing organism. These chemical agents may be in food additives, drinking water, industrial waste, or pesticides. Perhaps the early evidence of embryonic effect of chemicals in mammals dates back to Hale (1937) when he discovered that vitamin A deficiency affects the normal development of the eye in pigs. That was probably the first sign of discovery that the placenta is not protective and by changing the conditions of the mother, the child is also affected. Between the 1950s and 1970s a drug called aminopterin, which is an analog of amino acid, was used for ending unwanted pregnancies. It was later discovered that in several cases when the abortion was unsuccessful, the drug caused malformation and other developmental effects in the children. The most disastrous evidence of developmental toxicity is the thalidomide tragedy. In the early 1950s and early 1960s the drug was used by millions of pregnant women to offset morning sickness. In that period, more than 10,000 children were born in 46 countries around the world with limb defects. The strange and surprising fact about the thalidomide was that there was no evidence that the drug caused any adverse effects to adult humans, but it clearly was toxic to the embryo. These and other similar events brought awareness of the importance of the study of developmental toxicants. Several thousands of experiments were performed after the 1970s to try and identify chemicals that are teratogenic and can cause adverse effects to the developing fetus.

In this chapter, we discuss the general design of developmental toxicity experiments and describe the expected outcomes from such experiments. These outcomes are used for risk assessment of developmental toxins and our attention will particularly be focused on the critical issue of dose-response modeling. This stage involves the determination of a mathematical relationship between the mother's exposure level and the severity of the effects on the offspring. The dose-response models are utilized for extrapolation and estimation of risk at the human exposure levels. We show how accurate statistical models can play a crucial role and lead to more reliable estimates of risk.

5.2 Design of Reproductive and Developmental Toxicity Experiments

The first problem that the researchers studying reproductive and developmental effects faced was how to design the experiments and in designing these studies, the primary question was the species that could be used as experimental units. Fortunately, although there are major differences in the adult mammals, their embryos behave very similarly and several different mammals were considered as candidates. Today there are many strains of laboratory mice, rats, and rabbits that are considered suitable for developmental toxicity experimentation. The advantage with these animals is that they have short pregnancy terms and give birth to several pups each time giving way to more efficient experiments. Next was the question of dosing. Generally, humans are exposed to a very low dosage level and in order to elicit the effect on a limited number of laboratory animals, dose levels much higher than human exposure levels had to be considered. In order to derive appropriate doses, in many cases some dose finding experiments are performed to determine levels that would lead to most informative results. The general design of developmental toxicity experiments consists of using between 20 and 25 pregnant female animals. The animals are randomly assigned and exposed to three to four dose levels of the chemical agent, including the placebo controls. The exposure usually occurs during a critical time of the gestation period. The route of exposure could be through food, drink, injection, or gavage and the animals are carefully kept under a controlled environment. To examine the physical effects of the agent, the animals are generally sacrificed just before term and the uterine content is examined for a variety of effects. For more on the design of developmental toxicity experiments, we refer to Tyl and Marr (2006). If the goal of the experiment is to study the behavioral effects of a substance, the animals are kept through the full pregnancy period. We discuss the statistical models for developmental neurotoxicity studies in Chapter 7.

The studies of reproductive and developmental effects are generally classified into three segments. Segment I is the study of fertility and reproductive ability. Segment II consists of the effect of exposure during structural development, and Segment III concerns the perinatal and postnatal studies, that is the effect of the toxin later in life. Most of the statistical modeling work is in Segment II. In such studies, there may also be a variety of different outcomes that could be of interest in developmental toxicity experiments. First is the number of fetuses in each litter. Next, some fetuses in the uterine could be resorbed, that is when the skeleton is destructed, making the fetus useless. Also some fetuses may be dead. Usually there is a higher chance of death or resorption in higher dose levels. If the percentage of occurrence of death or resorption is too high, it generally means that the experimental dosage level is too high and uninformative (Figure 5.1). In these cases, perhaps lower exposure levels should be considered. A fetus that is not dead or resorbed is called a viable fetus. Outcomes that could be of interest in live fetuses include weight, any sign of birth or skeletal defects, occurrence of malformations, and possible growth retardation. Sometimes some visceral outcomes on specific internal organs of the body such as lung, heart, liver, pancreas, and intestine may be of interest. Statistical models have focused mainly on developmental effects of toxicants and not so much on the reproductive system. Note that

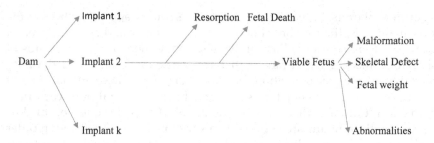

FIGURE 5.1
Outcome structure in developmental toxicity studies.

the data from developmental studies could be dichotomous, as in affected and unaffected fetuses. They could be multivariate, which is when several outcomes are considered. The data could be continuous when some measure on an interval scale such as fetal weight or organ weight is considered. The data may also be in terms of count. The guidelines for developmental toxicity testing and risk assessment were published by the US Environmental Protection Agency (EPA, 1991). Most of the early attention in developmental toxicity studies was focused on dichotomous responses. The outcome of a developmental toxicity study is dichotomous when the number of affected fetuses is recorded for each litter. Here, affected fetus refers to a fetus that is either resorbed or dead, or shows a sign of malformation and abnormality. In some studies, only viable fetuses are considered and an affected fetus would be a fetus with a sign of abnormality.

Example 5.1

A large-scale study was conducted at the National Center for Toxicological Research (NCTR), US Food and Drug Administration (US FDA), in which several strains of mice were exposed to the herbicide 2,4,5-trichlorophenoxyacetic acid (2,4,5-T) during days 6–14 of pregnancy (Holsen et al., 1992). The experiment consisted of exposing 7500 pregnant dams in eight strains of mice to between six and eight dose levels of the compound. Data were collected on more than 77,000 offspring. The significant developmental effects observed were reduction in fetal weight and rise in incidence of cleft palate with increasing dose. It was noted that among untreated (control) animals, fetuses with cleft palate weighed less than fetuses without cleft palate. Table 5.1 gives the malformation summary of the data.

It is also to be noted that a fundamental feature of developmental toxicity studies is the fact that fetuses in the same litter appear to behave more similarly than fetuses from different litters. Existence of this so-called intralitter correlation, which causes overdispersion and the extra binomial variation, has added more difficulty in modeling the outcomes. In fact ignoring this litter effect can lead to severe error and wrong understanding of the results. Thus a major question in statistical modeling of outcomes from developmental toxicity studies has been how to

TABLE 5.1

Average Percentage of Fetuses with Cleft Palate per Litter in Strains of Mice

Strain							
CD-1							
	Dose mg/kg/d	0	30	45	60	75	90
	No. Litters	73	87	93	66	36	12
	% Affected	0.2	2.9	13.7	35.8	64.1	82.1
BALB/c							
	Dose mg/kg/d	0	30	45	60	75	90
	No. Litters	45	33	45	31	28	13
	% Affected	0.4	0	1.3	5.1	7.3	15.4
C3H He							
	Dose mg/kg/d	0	30	45	60	75	90
	No. Litters	129	115	83	96	11	19
	% Affected	0.6	0.9	3.1	8.5	3.8	60.3
C57BL/6							
	Dose mg/kg/d	0	15	30	45	60	75	90
	No. Litters	71	13	54	74	69	37	29
	% Affected	0.7	1.5	2.8	4.5	13.9	23.5	42.5
A/J							
	Dose mg/kg/d	0	15	20	25	30	45	60
	No. Litters	89	86	56	40	76	33	9
	% Affected	10.2	19.5	18.2	29.7	57.6	74.2	100

Data from Holson et al. (1992).

account for the litter effect. Several methods have been introduced for modeling and incorporating the intralitter correlation with each method impacting the risk assessment and determination of the benchmark dose (BMD). We will discuss many of these methods in the next few sections of this chapter.

5.3 Models for Incorporating the Litter Effect

5.3.1 The Beta-Binomial Model and Likelihood Models

Perhaps the earliest attempt to model dichotomous data from developmental toxicity studies is by Williams (1975). Suppose that the experiment consists of a control and g non-zero doses:

$$0 = d_0 < d_1 < \cdots < d_g$$

Assume that m_i ($i = 0, 1, \ldots g$) pregnant female animals are exposed to the dose d_i of a developmental toxicant according to some exposure regimen.

Let n_{ij} and X_{ij}; $j = 1, 2, ..., m_i$ be respectively the litter size and the number of affected fetuses of the jth animal in the ith dose group. Then if we denote by p_{ij} the probability of response (being affected) in the jth animal in the ith dose group, then we have

$$P(X_{ij} = x_{ij} \mid p_{ij}) = \binom{n_{ij}}{x_{ij}} p_{ij}^{x_{ij}} (1 - p_{ij})^{n_{ij} - x_{ij}} \quad x_{ij} = 0, 1, ..., n_{ij}.$$

Now, as mentioned before, because of the existence of the intralitter correlation, the probability of response varies from one litter to another. If we assume that the variation in the response probability can be expressed by a beta distribution, that is if

$$g(p_{ij}) = \frac{1}{B(\alpha_i, \beta_i)} p_{ij}^{\alpha_i - 1} (1 - p_{ij})^{\beta_i - 1} \quad \alpha_i, \beta_i > 0$$

where $B(\alpha_i, \beta_i)$ is the beta function given by

$$B(\alpha_i, \beta_i) = \frac{\Gamma(\alpha_i + \beta_i)}{\Gamma(\alpha_i)\Gamma(\beta_i)},$$

then the unconditional distribution of X_{ij} is given by

$$P(X_{ij} = x_{ij}) = \int_0^1 P(X_{ij} = x_{ij} \mid p_{ij}) . g(p_{ij}) dp_{ij}$$

$$= \binom{n_{ij}}{x_{ij}} \frac{B\left(\alpha_i + x_{ij}, n_{ij} + \beta_i - x_{ij}\right)}{B(\alpha_i, \beta_i)} \quad x_{ij} = 0, 1, ..., n_{ij}$$

which is the familiar beta-binomial distribution and is often used to fit over-dispersed data.

The properties of the beta-binomial distribution are well known. For example, we can easily prove that the mean and variance of X_{ij} are given by

$$E(X_{ij}) = n_{ij} \frac{\alpha_i}{\alpha_i + \beta_i}$$

and

$$V(X_{ij}) = n_{ij} \frac{\alpha_i \beta_i \left(\alpha_i + \beta_i + n_{ij}\right)}{\left(\alpha_i + \beta_i\right)^2 \left(\alpha_i + \beta_i + 1\right)}.$$

A reparameterization of the beta-binomial distribution, which is more amenable to developmental toxicity data can also be obtained by letting

$\mu_i = \dfrac{\alpha_i}{\alpha_i + \beta_i}$ and $\theta_i = \dfrac{1}{\alpha_i + \beta_i}$. Note that μ_i can be interpreted as the mean of the beta distribution that describes the litter effect at the ith dosage level. The variance of that distribution is given by

$$\frac{\alpha_i \beta_i (\alpha_i + \beta_i + n_{ij})}{(\alpha_i + \beta_i)^2 (\alpha_i + \beta_i + 1)} = \mu_i (1 - \mu_i) \frac{\theta_i}{1 + \theta_i}.$$

Thus given μ_i, the parameter θ_i determines the shape of the distribution and when $\theta_i = 0$, the pure binomial case is derived. Note also that with the new parameterization, we have

$$E(X_{ij}) = n_{ij} \mu_i \tag{5.1}$$

and

$$V(X_{ij}) = n_{ij} \mu_i (1 - \mu_i) \left\{ 1 + (n_{ij} - 1) \frac{\theta_i}{1 + \theta_i} \right\} \tag{5.2}$$

The expression for the variance clearly displays the overdispersion. The extra variation over the pure binomial is expressed through the term $n_{ij} \mu_i (1 - \mu_i)(n_{ij} - 1) \frac{\theta_i}{1 + \theta_i}$. With this model, the fetal correlation, assumed to be the same for all litters across the dose group is given by

$$\varphi_i = \frac{\theta_i}{1 + \theta_i} = \frac{1}{(\alpha_i + \beta_i + 1)} \quad i = 0, 1, \ldots, g.$$

Note that when the fetal correlation is zero, the pure binomial case is obtained. The main advantage of the beta-binomial model is its simplicity and ease of use. But major drawbacks are the strong parametric assumptions and also the fact that it only accommodates positive correlation.

The log-likelihood for the beta-binomial model is

$$\log L = \sum_{i=0}^{g} \sum_{j=1}^{m_i} \left\{ \sum_{k=0}^{x_{ij}-1} \log(\mu_i + k\theta_i) + \sum_{k=0}^{n_{ij}-x_{ij}-1} \log(1 - \mu_i + k\theta_i) - \sum_{k=0}^{n_{ij}-1} (1 + k\theta_i) \right\} \tag{5.3}$$

and by maximizing this function, the parameter estimates can be obtained. Maximizing the likelihood function for the beta-binomial model is now a standard problem and many modern statistical software have a function for determining the parameter estimates. For example R can routinely calculate the maximum likelihood estimates of the parameters. Williams (1975) used the beta-binomial model to detect the response differences between a group of untreated animals and a group of animals treated with a chemical during pregnancy through lactation. No specific dose-response model was used. However, to estimate the probability of an adverse effect at a given dose, a parametric model is assumed for μ_i and parameters are estimated. A popular class of parametric models in toxicology is

$$\mu(d) = F(\gamma_0 + \gamma_1 d)$$

where $F(.)$ is some appropriate cumulative distribution function. The simplest model in this class is the one-hit model or the cumulative distribution function (cdf) of the exponential distribution

$$\mu(d) = 1 - \exp[-(\gamma_0 + \gamma_1 d)] \tag{5.4}$$

This model assumes that the adverse effect occurs once a target tissue is hit by a single biologically effective dose. By substituting this model in (5.3) and differentiating with respect to the unknown parameters, we can derive the likelihood equations and solve to find the parameter estimates (see Exercise 1 at the end of this chapter). Logistic and probit models are also frequently used to describe dose-response relationships. However, the earliest attempt to develop a model specifically for use in developmental toxicity studies was by Rai and Van Ryzin (1985). They proposed a modified version of the one-hit model incorporating the litter size as a parameter

$$\mu(d) = \exp\left[-n_{ij}\left(\delta_0 + \delta_1 d\right)\right]\left[1 - \exp\left[-(\gamma_0 + \gamma_1 d)\right]\right].$$

This model assumes that the conditional distribution of risk given the litter size is binomial. The advantage of the model is that it assumes that probability of response is a decreasing function of the litter size; i.e. it is based on the argument that litters with larger sizes are likely to be healthier. Using the beta-binomial model, Chen and Kodell (1989) applied the Weibull model and Kupper et al. (1986) proposed an extension of the Weibull model. Faustman et al. (1989) propose a model for estimating the individual pup risk at dose d as

$$\mu(d) = \left[1 - \exp[-(\gamma_0 + \gamma_1 d)]\right]\exp\left[-\theta_1 \exp(-\theta_2)\right]$$
$$\left[1 - \exp[-(\delta_0 + \delta_1 d)]\right]$$

where $\gamma_0, \gamma_1, \theta_1, \theta_2, \delta_0,$ and δ_1 are model parameters to be estimated.

5.3.2 The Quasi-Likelihood Models

Although the beta-binomial model is an appealing and flexible approach to incorporate the litter effect, it relies on strong parametric assumptions and more importantly, it only incorporates positive correlation. A few extensions of the beta-binomial model have been introduced that account for negative correlations as well. For example, Prentice (1986) introduces a reparameterization of the beta-binomial model and shows that a negative correlation is possible. Through an extensive simulation study, Kupper et al. (1986) show that when the intralitter correlations are dose-dependent, the assumption of a constant intralitter correlation induced by the beta-binomial distribution could lead to severe bias in parameter estimation. They point out that the beta-binomial distribution is used mainly for mathematical convenience rather than biological principles. Because of the shortcomings of the beta-binomial distribution, methods in which the impact of the litter effect is reduced have become more popular. One such technique is the quasi-likelihood method proposed first by Wedderburn (1974). The main advantage of this method is that it does not require a fully specified distribution for the response variable. As described by Wedderburn (1974), quasi-likelihood models generalize the idea of weighted least squares and method of moments by setting weighted sums of the observed data equal to their expected values under the assumed model for the mean. Therefore, the quasi-likelihood method does not rely on strong distributional assumption and only the specification of the mean as a function of the parameters and the variance as a function of the mean are required. Generally, the quasi-likelihood estimate for the vector of parameters β is obtained by solving the set of quasi-likelihood equations

$$U(\beta) = D^T V^{-1}(X - \mu)$$

called the quasi-score function. D is the matrix of the first derivatives of the mean function $\mu(\beta)$ with respect to components of β, X is the vector of observations, and V is the diagonal matrix of variances. The covariance matrix of $\hat{\beta}$ is given by

$$V(\hat{\beta}) = \left(D^T V^{-1} D \right)^{-1}.$$

The application of the quasi-likelihood method in developmental toxicity was first presented by Williams (1982). Taking the same moments as for the beta-binomial model, Williams (1982) showed that consistent estimators of the parameters characterizing the mean response rate $\mu(d)$ can be found by solving the set of equations

$$\sum_{i=1}^{g}\sum_{j=1}^{m_i} \frac{\partial\mu(d_i)}{\partial\beta} \cdot \frac{(x_{ij} - n_{ij}\,\mu(d_i))}{n_{ij}\mu(d_i)(1 - \mu(d_i))[1 + \phi(n_{ij} - 1)]} = 0, \tag{5.5}$$

where β represents the parameters of $\mu(d_i)$ and ϕ is the intralitter correlation. Quasi-likelihood models can be fitted using a straightforward extension of the algorithms used to fit generalized linear models. However, the criticism regarding the constant intralitter correlation still exists.

5.3.3 Generalized Estimating Equations

The introduction of the generalized estimating equations (GEEs) first by Liang and Zeger (1986) and later by Liang et al. (1992) to account for correlation in clustered data stimulated a notable interest among researchers. The GEE approach is essentially an extension of the generalized linear models (GLMs) where within cluster covariance structure is treated as a nuisance parameter. An important feature of the methodology is the inclusion of an empirical, robust estimator that provides correct inferences for the model parameters and their variance, even in cases where the assumed variance structure is misspecified. The generalized estimating equations can be written in vector form as:

$$\sum_{i=1}^{g}\sum_{j=1}^{n_i} D_{ij}^{T} V_{ij}^{-1} (X_{ij} - \mu_i) = 0 \qquad (5.6)$$

where $X_{ij} = (X_{ij1},\ldots,X_{ijn_{ij}})^{T}$ is the vector of fetus-specific indicators x_{ijk} of the jth litter of the ith dose group, $\mu_i = \mu(d_i)$ is the mean vector of outcomes x_{ij}, D_{ij} represents $n_{ij} \times p$ design matrix $\dfrac{\partial \mu}{\partial \beta}$ where, as before, β represents the p-vector of unknown parameters characterizing $\mu(d_i)$, and V_{ij} is the $n_{ij} \times n_{ij}$ assumed covariance matrix of X_{ij}. Writing V_{ij} as

$$V_{ij} = A_{ij}^{1/2} R(\alpha) A_{ij}^{1/2}$$

where $R(\alpha)$ is a suitable correlation matrix for x_{ij} and A_{ij} is the diagonal matrix

$$A_{ij} = \text{diag} \left[\mu(d_i)(1 - \mu(d_i)) \right],$$

Liang and Zeger (1986) construct a robust empirical variance estimator to adjust for misspecification in the assumed model. If R and A are chosen to have reasonably sensible structures for the data, the GEE approach will yield highly efficient estimates of the parameters. Ryan (1992b), Catalano and Ryan (1992), and Razzaghi (1999) describe the applications of the GEE method in developmental toxicology. Liang and Hanfelt (1994) show through extensive simulation studies that the impact of the intralitter correlation is minimal when the quasi-likelihood method is used. They recommend the use of this method with a common overdispersion parameter for analyzing developmental toxicity data. Several extensions of the GEE methodology were subsequently introduced for which the interested reader is referred to Hall and Severini (1998).

5.4 Dose-Response Models

The primary purpose of dose-response models is to provide a descriptive summary of experimental data as a function of dose. Because of the complexity of the developmental process, there is little understanding of how dose affects the developing fetus and consequently the probability of an adverse effect. Gaylor and Razzaghi (1992) argue that the quantitative evaluation and interpretation of endpoints in developmental toxicology depends on an understanding of the biological events leading to the endpoints observed, the relationship among endpoints, and their relationship to dose and maternal toxicity. If a model is based on biological principles, it can characterize the potential mechanistic process of a developmental defect and it may provide a mathematical framework to test if experimental data are consistent with hypothesized mechanisms. Assuming that the development of an offspring from conception to birth is a sequence of n independent events and in order for a fetus to be normal, all stages of the sequence of this process must develop normally, the probability of giving birth to an abnormal fetus is

$$P(D) = 1 - \prod_{i=1}^{n} (1 - P_i)$$

where D is the administered dose and P_i is the probability of abnormal development of the ith stage for $i = 1, \ldots, n$. In its simplest case, assume that a toxin may only affect one stage of the process. Thus if it is known that stage j is affected, then all other stages behave the same as the background. Therefore $P(D)$ can be expressed as

$$P(D) = 1 - (1 - P_j(D)) \prod_{i \neq j} (1 - P_i(D)) = 1 - \frac{(1 - P_j(D))}{(1 - P_j(0))} (1 - P(0)) \qquad (5.7)$$

where $P_i(D)$ is the probability of abnormal development of the ith stage at dose D and $P(0)$ denotes the probability of an abnormal development in the untreated control animals.

For illustration, consider a stage at which cell replicationis required to reach a critical number of cells for normal growth to proceed. Assuming that the relative probabilities of successful cell growth for normal development at the jth stage is equal to the ratio of the number of cells produced, that is

$$\frac{(1 - P_j(D))}{(1 - P_j(0))} = \frac{N(t, D) - N_0}{N(t, 0) - N_0}$$

Assuming further that cell growth can be modeled as exponential in time, then (5.7) is reduced to

$$P(D) = 1 - \frac{1 - e^{\beta(D)t}}{1 - e^{\beta(0)t}}(1 - P(0))$$

To describe quantal bioassay data based on counts of individuals possessing a characteristic such as malformation or fetal death, Gart et al. (1986) suggested a class of dose-response models

$$\mu(d) = F(\gamma_0 + \gamma_1 d) \tag{5.8}$$

where F is a cumulative distribution function. The probit and the logistic models are some examples of dose-response models that belong to this class. A modification of the class of models (5.5) which is more flexible can be represented by

$$\mu(d) = F(\gamma_0 + \gamma_1 d^s) \quad s > 0. \tag{5.9}$$

The Weibull model

$$\mu(d) = 1 - \exp(\gamma_0 + \gamma_1 d^s)$$

is an example of this latter class which has gained widespread popularity in dose-response modeling, although the logistic and probit models can easily be adapted to belong to the class (5.6) by adding a power parameter to the dose. The class of models (5.5) and (5.6) have been used by several investigators to model developmental toxicity data with varying degrees of success. For example, Kupper et al. (1986) use the logistic distribution, Chen and Kodell (1989) use the Weibull distribution, and Ryan (1992a) uses the modified probit model. As observed by Gaylor (1994), the fact is that most choices in the families (5.5) and (5.6) provide good fits to observed data in the experimental range, but may diverge dramatically at lower doses to which humans are exposed. If there is some biologic basis for the choice of the dose-response model, then extrapolation of results outside of the experimental dose may be more meaningful. Based on this argument, Razzaghi (1999) introduces the mixture models for dose-response modeling. Specifically, the class of models

$$\mu(d) = \theta F(\gamma_{01} + \gamma_{11}d) + (1 - \theta)F(\gamma_{02} + \gamma_{12}d) \tag{5.10}$$

is proposed. It is argued that the main advantage of this new class of models is that not only is the model more flexible, and therefore provides a better fit for the data, but it also accounts for any non-homogeneity in the population. For example, in many experiments with laboratory animals it frequently happens that some subjects are more susceptible to a dose of a chemical. A mixture model has been used to describe the distribution of responses for such populations; see for example Salsburg (1986), Boos and Brownie (1986), Conover and Salsburg (1988), and Razzaghi and Nanthakumar (1992, 1994). Boos and Brownie (1991) describe several bioassay experiments in which some animals are not affected by treatment and discuss the problem of dose-response modeling for quantitative responses in the presence of nonresponders in the treatment group and used a mixture of two normal distributions to represent the dose-response model. An additional advantage of the mixture is that it allows for bimodality and hence accounts for any major non-homogeneity such as susceptibility of animals to different chemicals. A mixture of two logistic distributions with a common background parameter $\gamma_{01} = \gamma_{02} = \gamma_0$,

$$\mu(d) = \theta\left\{1 + \exp\left[-\left(\gamma_0 + \gamma_{11}d\right)\right]\right\}^{-1} + (1-\theta)\left\{1 + \exp\left[-\left(\gamma_0 + \gamma_{12}d\right)\right]\right\}^{-1} \quad (5.11)$$

was used. Molenberghs and Ryan (1999) proposed a conditional likelihood-based model based on a multivariate exponential family that describes the probability of an outcome given the values for the other outcomes. They proposed the distribution of X_{ij}, that is, the number of viable fetuses with positive response from the jth dam in the ith dose group as

$$P\left(x_{ij}; \theta_i, \delta_i, n_{ij}\right) = A\left(\theta_i, \delta_i\right)\exp\left[\theta_{ij}x_{ij} - \delta_{ij}x_{ij}\left(n_{ij} - x_{ij}\right)\right]$$

where $\theta_i = \left(\theta_{i1}, \ldots, \theta_{ik}\right)^T$ is the vector of parameters and $\delta_i = \left(\delta_{i1}, \ldots, \delta_{ik}\right)^T$ describes the intralitter correlations. Faes et al. (2006) demonstrate the use of fractional polynomials as dose-response models and show through simulation that using a linear predictor in the dose-response model can yield unrealistically low or unreliable safe doses, even when the probability model is well specified.

5.5 Risk Assessment for Developmental Toxicity Experiments

Traditionally, as in all non-cancer endpoints, the process of risk assessment in developmental toxicity experiments involved the use of the no-observed-adverse-effect level (NOAEL), which is defined as the highest experimental dosage level that produces no statistically or biologically adverse effect compared to the control animals. To determine a reference dose (RfD), the NOAEL is divided by a series of safety factors to account for uncertainty in extrapolating from animals to humans, variability among human sensitivities to toxic agents, and inadequate data. It is, however, clearly shown now (see e.g. Leisenring and Ryan (1992) and Gaylor (1994)) that the NOAEL approach is unreliable, ill-defined, subjective, dependent on dose-spacing, and does not reward better experimentation. As described in Chapter 2, model-based approaches leading to determination of upper confidence limits on risk and benchmark doses have become fairly standard and universally accepted methods for risk assessment. Risk $P(d)$, or the assumed dose-response model, is defined as the probability of observing an adverse effect, for example, mal-formation. In developmental toxicity studies, however, this definition is not quite clear. The question is, When and, with the observation of how many fetuses with at least one malformation within a litter, do we call the response adverse? Declerk et al. (2000) argue that from a biological perspective, it is important to consider the overall health of the entire litter when modeling risk and therefore and the risk function $P(d)$ should be defined as the prob-ability of observing a malformation in at least one fetus within the litter as the result of exposure to a developmental toxicant at dosage d. Based on this definition, the measure of risk as defined in Chapter 2 can be defined as either the additional risk

$$\pi(d) = P(d) - P(0)$$

which is the probability of observing an adverse effect at exposure level d over the background risk or

$$\pi(d) = \frac{P(d) - P(0)}{1 - P(0)}$$

called the excess risk. Note that as $P(0)$ increases, the same absolute change in the risk function results in greater changes in the excess risk and thus as noted by Crump (1984), the additional risk places equal weight at all levels of the background risk while the excess places greater weight on outcomes with higher background risk. The two measures of risk become very similar when the background risk is small.

Model-based approaches for risk assessment involve fitting a dose-response model to the data and using the model for extrapolation and esti-mating the risk at a given dose. However, because of the inherent variability and the uncertainty due to the choice of the dose-response model, often the upper confidence limit on risk is utilized. There are a few approaches to determine the upper confidence limit for risk. Here, we present three popu-lar approaches:

- One approach is to use the asymptotic distribution of the likelihood ratio statistic. In this case, the process of risk assessment begins by calculating the maximum likelihood estimates of the model param-eters. Let L_{max} be the value of the likelihood function evaluated at the parameter estimates. Then the $100\,(1 - \alpha)\%$ upper confidence limit for risk at dose D is the value of $\pi(D)$ that satisfies

$$2\left[\log L_{\max} - \log L\left[\pi(D)\right]\right] = \chi^2_{1,2\alpha}$$

where $L[\pi(D)]$ is the maximum of the likelihood function subject to the risk restriction

$$\hat{\pi}(D) = R$$

and $\chi^2_{1,2\alpha}$ is the 100 $(1 - \alpha)$th percentile of the χ^2 distribution with 1 degree of freedom. This approach is computationally complex and could be rather unstable.

- A simpler alternative approach that has found more popularity is to use the asymptotic normality of the parameter estimates and use

$$\pi^*(D) = \hat{\pi}(D) + \Phi^{-1}(1-\alpha)\left[\widehat{\mathrm{Var}}\left(\hat{\pi}(D)\right)\right]^{\frac{1}{2}}$$

where $\Phi^{-1}(1-\alpha)$ is the 100 $(1 - \alpha)th$ percentile of the standard normal distribution and $\widehat{\mathrm{Var}}(\hat{\pi}(D))$ can be estimated from

$$\widehat{\mathrm{Var}}(\hat{\pi}(D)) = \left(\frac{\partial \hat{\pi}(D)}{\partial \beta}\right)^T \hat{\Sigma}(\hat{\beta})\left(\frac{\partial \hat{\pi}(D)}{\partial \beta}\right) \tag{5.12}$$

in which β denotes the vector of unknown parameters and $\hat{\Sigma}(\hat{\beta})$ is the estimate of the variance–covariance matrix of parameter estimates.

- A third approach, which has not been used very often, is bootstrapping. Bootstrap samples are generated from the given data and the associated risk is calculated. The process is repeated a large number of times e.g. 10,000 times and the desired percentage is selected. For example if 95% upper confidence limit on risk is desired and 10,000 bootstrap samples are generated, the 95th percentile of the 10,000 risk values is selected. For more on bootstrap approach refer to Buckley et al. (2009) and Bailer and Smith (1994).

Alternatively, in risk assessment, it is often the estimate of the safe exposure level or what is commonly known as the BMD that is of interest. The BMD is the dosage level that gives rise to a nominal low risk w usually chosen to be between 1% and 10% often referred to as benchmark risk (BMR). A dose-response model is fitted to the data and is used for extrapolating to low human exposure levels. Thus if w is a pre-specified level of increased risk, then $\mathrm{BMD}_w = \pi^{-1}(w)$. Therefore BMD is determined by inverting the assumed risk equation. Specifically, if the additional risk is used as the measure of risk, that is,

$$\text{BMD}_w = P^{-1}[P(0) + w]$$

and if for example a dose-response function from the class $F(\gamma_0 + \gamma_1 d)$ such as the logistic or the probit models is used, then

$$\text{BMD}_w = F^{-1}[F(\gamma_0) + w]$$

where in practice the estimate of γ_0 is used to determine the BMD. Once again, because of the inherent variability of the estimates, often a 100 (1 − α)% BMD lower confidence limit (BMDL) is determined. Thus

$$\widehat{\text{BMDL}_w} = \widehat{\text{BMD}_w} - \Phi^{-1}(1 - \alpha)\left[\widehat{\text{Var}\left(\widehat{\text{BMD}_w}\right)}\right]^{\frac{1}{2}} \tag{5.13}$$

where $\left[\widehat{\text{Var}\left(\widehat{\text{BMD}}\right)}\right]$ can be determined from

$$\widehat{\text{Var}\left(\widehat{\text{BMD}}\right)} = \left(\frac{\partial \text{BMD}}{\partial \beta}\right)^{T} \hat{\Sigma}(\hat{\beta})\left(\frac{\partial \text{BMD}}{\partial \beta}\right)$$

or alternatively, as is very common in practice, bootstrapping can be applied to find BMD_w. Thus bootstrap samples are generated from the data and the value of BMD_w is estimated for each sample. The process is repeated a large number of times, and the αth percentile is selected as the estimate of $BMDL_w$.

An alternative approach to estimate a safe exposure level was introduced by Kimmel and Gaylor (1988) and Williams and Ryan (1996). The method is based on first finding the upper confidence limit on risk and then solving for the dose that makes this upper bound equal to a pre-specified risk. The term lower effective dose (LED) is used for this dosage level. Therefore, for a given risk of say w, the LED is given by the solution of

$$\hat{\pi}(d) + \Phi^{-1}(1 - \alpha)\left[\widehat{\text{Var}\left(\hat{\pi}(d)\right)}\right]^{\frac{1}{2}} = w$$

where $\widehat{\text{Var}}(\hat{\pi}(d))$ is determined from (5.12). The benchmark dose methodology, however, has by far found popularity and has become the standard for risk assessment and determination of the safe exposure level.

Example 5.2

In this example we illustrate the application of the quasi-likelihood method and the GEE approach using the data set described in Example 5.1, where the developmental effect of the chemical 2,4,5-T was examined on several strains of mice. The outcome of interest in the data is the incidence of cleft palate in the fetuses. For our purpose, we use the data on CD-1 mice and for the dose-response model we use a mixture of two

TABLE 5.2

Developmental Effect of 2,4,5-Trichlorophenoxyacetic acid (2,4,5-T) in CD-1 Mice

Dose	Dams	Implants	Live	Malformed
0	71	802	745	2
30	87	956	855	24
45	93	1073	891	109
60	66	712	568	152
75	36	402	268	158
90	12	121	69	57

Implants = live + dead/resorbed.

logistic distributions given in (5.13). As described before, the advantage of using a mixture is that not only is it a more flexible distribution and probably provides a more adequate fit to the data, but it also accounts for any non-homogeneity such as susceptibility to the drug in the subjects. For more details, we refer to Razzaghi (1999). Table 5.2, partially reproduced from Table 1 in Razzaghi (1999), but also providing information about the number of implants, the number of live fetuses, as well as the number of pups with cleft palate malformation, provides a summary of the data for the CD-1 strain of mice.

Using a common intralitter correlation coefficient, the GEE method with mean and variance induced by the beta-binomial distribution given in (5.1) and (5.2) respectively leads to a set of four nonlinear equations in the four unknown parameters θ, γ_0, γ_1, and γ_1 that need to be solved using a root finding procedure. To estimate the overdispersion parameter ϕ, a two-step procedure was used where an initial value of ϕ is first selected and used to derive the estimates for the four model parameters. The estimates are then substituted in the equation for the overdispersion parameter given by

$$\hat{\phi} = \frac{1}{\sum_{i=0}^{g} m_i - 4} \sum_{i=0}^{g} \sum_{j=1}^{m_i} \frac{\left(y_{ij} - \widehat{\mu}(d_i)\right)^2}{n_{ij}\widehat{\mu}(d_i)\left(1 - \widehat{\mu}(d_i)\right)\left(\phi^{-1} + n_{ij} - 1\right)}$$

where

$$\widehat{\mu}(d_i) = \hat{\theta}\left\{1 + \exp\left[-\left(\hat{\gamma}_0 + \hat{\gamma}_{11}d\right)\right]\right\}^{-1} + (1 - \hat{\theta})\left\{1 + \exp\left[-\left(\hat{\gamma}_0 + \hat{\gamma}_{12}d\right)\right]\right\}^{-1}.$$

The new value of $\hat{\phi}$ is substituted in the estimating equations and solved. The process is continued until a desired level of convergence is reached. Using the additional risk

$$\hat{\pi}(D) = \widehat{\mu}(D) - \widehat{\mu}(0)$$

as the measure of risk at the given exposure D, the benchmark doses are calculated by using equation (5.13). Table 5.3 gives the 95% BMDL for selected values of additional risk.

TABLE 5.3

Benchmark Doses for CD-1 Mice Exposed to 2,4,5-Trichlorophenoxyacetic acid (2,4,5-T)

Risk	Benchmark Dose (BMD)	SD	BMD Lower Confidence Limit (BMDL)
0.05	34.13	12.77	13.12
0.06	35.71	11.89	16.81
0.07	37.06	10.44	19.89
0.08	38.26	9.59	22.51
0.09	39.33	8.83	24.77
1	40.3	8.25	26.74

Example 5.3

Chen and Kodell (1989) describe an experiment regarding the developmental toxicity effect of the chemical diethylhexyl phthalate (DEHP) in mice. The experiment consisted of five dose levels, and the developmental effect was considered to be death/resorption or malformation. Table 5.2 provides a summary of the data. Chen and Kodell (1989) used the Weibull dose-response model along with the beta-binomial model to find the maximum likelihood estimates of the parameters.

Using a likelihood ratio test, they show that the beta-binomial model provides a more adequate description of the data than the pure binomial model; i.e. the data exhibit extra binomial variation. They estimate the so-called effective dose BMD_{01}, the dose affecting 1% of the animals to be 40.7 mg/kg of body weight with a lower confidence limit of $BMDL_{01} =$ 28.5 mg/kg body weight. Ryan (2002) used the same data set and applied the modified probit model

$$\mu(d) = \Phi(\gamma_0 + \gamma_1 d^s) \quad s > 0$$

which is a member of the general class equation (5.6). The generalized estimating equation technique with a constant correlation matrix is applied for the incorporation of the intralitter correlation and estimation of the parameters. The estimated benchmark dose with 1% risk was found to be $BMD_{01} = 28$ mg/kg body weight, which is much lower than the result of Chen and Kodell (1989). The difference is apparently due to the fact that the probit model rises slightly more rapidly than the Weibull model. The 95% BMDL was found to be $BMDL_{01} = 13$ mg/kg body weight, which again is lower than Chen and Kodell's result. This example illustrates how the choice of the dose-response model can play a critical role in risk assessment.

5.6 Application of Model Averaging

As pointed out in the last chapter, a major problem in applying the model based risk assessment BMD methodology is that many dose-response models may fit the data quite well in the experimental dose range, but when extrapolated to the low levels, the BMD estimates can often vary by an order of magnitude. West et al. (2012) show that

> an uncomfortably high percentage of instances can occur where the true extra risk at the BMD lower confidence limit (BMDL) under a misspecified or incorrectly selected model can surpass the target BMR, exposing potential danger of traditional strategies for model selection when calculating BMDs and BMDLs.

To account for the uncertainty due to the choice of the dose-response model, model averaging has recently been introduced and utilized in a number of risk assessment problems. For example, Kang et al. (2000) and Moon et al. (2005) have applied the model averaging techniques for microbial risk assessment. Wheeler and Bailer (2007) and Wheeler and Bailer (2012) used model averaging for estimating the benchmark dose in cancer studies and found good results. In a later study, the same authors (Wheeler and Bailer, 2013) used a large amount of response data on trout to examine the performance of the model averaging procedure. Since observational data were available at the low-dose region, the authors were able to estimate the benchmark dose and compare with the observed risk at low doses. They found that the model averaging technique works well and that quantitative risk estimates based on model averaging is a "promising" alternative to linear extrapolation based on a single model. Piegorsch et al. (2013) also found that the model averaging approach to calculating the BMDs can result in improved risk estimation. In model averaging, rather than using a single mathematical relationship as dose-response model, several potential candidate models are considered and a weighted average of these models is utilized for extrapolation. An approach to model averaging that has found widespread popularity is the Bayesian Model Averaging (BMA).The advantage of BMA is that the weights are determined in such a way that they are proportional to the posterior probability that each model is correct given the observations. Therefore the weights show the extent of support in the data for each model. Whitney and Ryan (2009) consider the properties of the BMA technique in benchmark dose estimation and show that the derived estimates more accurately reflect uncertainty in the understanding of the effects of exposure on the occurrence of adverse responses. As pointed out by Hoetling et al. (1999), the BMA methodology is a coherent approach since we can express the desired quantities as a weighted average of model-specific quantities with the weights determined based on how much the data supports each model.

In developmental toxicity studies, the application of model averaging technique was recently demonstrated in Khorsheed and Razzaghi (2019). A full Bayesian approach is used to fit each dose-response model and the Markov Chain Monte Carlo (MCMC) method is utilized to determine the parameter estimates for each model. Thus if K candidate models are used in the model averaging procedure and if we denote the kth model by $P_k(d) = F_k(\gamma_{1k} + \gamma_{2k}d^h)$

for $k = 1, \ldots K$, then the joint posterior distribution of all model parameters, given the data is proportional to the product of the likelihood and the joint prior distribution of all the parameters, is

$$P(\gamma_1, \gamma_2 \mid X_1, \ldots Xg) \propto P(\gamma_1, \gamma_2) * L$$

where $\gamma_1 = (\gamma_{11}, \ldots \gamma_{1K})^T$, $\gamma_2 = (\gamma_{21}, \ldots \gamma_{2K})^T$, $X_i = (X_{i1}, \ldots, X_{im_i})^T$ $i = 1, \ldots, g$, $P(\gamma_1, \gamma_2)$ is the joint prior distribution of γ_1 and γ_2, and L is given by (6). Following Shao and Small (2011) we also assume that the prior distributions of parameters of all the dose-response models used for model averaging, i.e. $\gamma_{11}, \ldots, \gamma_{1k}$ and $\gamma_{21}, \ldots, \gamma_{2k}$, are independent and noninformative and that no parametric specification of the prior distribution is required. Once the parameter estimates are determined, the BMD for each model $P_j(d)$ for a given level of risk π^* can be derived from

$$\mathrm{BMD}_k = \left[\left\{ F_k^{-1}(\pi^*) - \gamma_{1j} \right\} \Big/ \gamma_{2k} \right]^{1/h} \qquad k = 1, \ldots, K.$$

By applying the MCMC methodology, we can generate a large sequence of parameter estimates for each model and determine the BMD. Using a measure of the center, e.g. the average, we have a point estimate for BMD and using the fifth percentile value, we have the corresponding BMDL.

In the BMA methodology, we begin by assuming that all the K models have *a priori* equal weights, that is

$$P(W_k) = \frac{1}{K} \quad k = 1, 2, \ldots, K.$$

Then, the final weights are determined by the posterior model probabilities, which by Bayes' theorem are given by

$$P(W_k \mid L) \propto K^{-1} P(L \mid W_k) \quad k = 1, 2, \ldots, K$$

where $P(L \mid W_k)$ represents the marginal distribution of the likelihood for each given model. Now, as explained in Whitney and Ryan (2009), the computation of the marginal distributions for calculation of the posterior model probabilities in the implementation of BMA requires solving an integral that is difficult to calculate except for very simple cases. Indeed, in most cases, especially data relating to environmental and epidemiological studies, derivation of closed form solutions is not feasible and the use of the Bayesian Information Criteria (BIC) to approximate the marginal distributions has successfully been adopted. Specifically, Raftery (1995) suggests the following approximation,

$$P(W_k \mid L) \propto \exp\left(-\frac{1}{2} \mathrm{BIC}(W_k) \right) \quad k = 1, \ldots, K$$

with

$$\text{BIC}(W_k) = -2\log(\max L \,|\, W_k) + a_k \cdot \log(n)$$

where a_k is the number of parameters for W_k, n is the sample size, and max L is the maximum of the likelihood function. Note that in developmental toxicity experiments, the sample size is the litter size. According to Wasserman (2000), this approximation works well in moderate sample sizes when the covariates are independent. Thus, the weights may be determined from

$$P(W_k \,|\, L) = \frac{\exp\left(-\frac{1}{2}\text{BIC}(W_k)\right)}{\sum_{r=1}^{K} \exp\left(-\frac{1}{2}\text{BIC}(W_r)\right)} \quad k = 1, \dots, K.$$

This procedure has been successfully applied in several applications with dichotomous responses, see for example Bailer et al. (2005) and Simmons et al. (2015). However Burnham et al. (2011) suggest replacing $\text{BIC}(W_k)$ by

$$\Delta(k) = \text{BIC}(W_k) - \min_{1 \leq r \leq K} \text{BIC}(W_r)$$

for calculating the weights. The advantage of this method is that the "Δ values are on a continuous scale of information and are interpretable regardless of the measurement scale and whether the data are continuous, discrete or categorical." Khorsheed and Razzaghi (2019) found that this approach was more computationally stable especially when $BIC(W_k)$ is relatively large. Using this approach on the data set utilized in Wheeler and Bailer (2007), the result was the same set of weights. For more information on advantages of using the Δ values, we also refer to Anderson (2008) and Burnham et al. (2002). Through some simulation studies, Khorsheed and Razzaghi (2019) show that the determined weights appear to have a consistency across the varying risk levels and that the BMA procedure works well in determining the associated weight for each model.

5.7 Statistical Models for Fetal Weight and Continuous Outcomes

As discussed previously, in many developmental toxicity studies, the outcome of interest is continuous. The fetal or organ weight are obvious examples of continuous outcomes. Although there is a large body of literature on the analysis of quantal data in toxicological experiments due to the fact that outcomes from carcinogenicity studies and incidence of malformation in developmental toxicity studies are binary in nature, there has been relatively less attention paid to analyzing quantitative data. However, due to the fact that the outcomes of neurotoxic effects and developmental neurotoxicity experiments are continuous, the analysis of non-quantal outcomes in the last 25 years has found widespread popularity. We will discuss the statistical models and analysis of continuous and non-quantal data from neurotoxicity and developmental neurotoxicity experiments in Chapters 6 and 7 respectively. In developmental toxicity experiments, however, Gaylor and Razzaghi (1992) observed that litters with higher proportion of malformation tend to have fetuses with lower weight.

Example 5.4

The developmental toxicity of ethylene glycol, a major industrial chemical used in antifreeze and coolant mixtures for motor vehicles, in hydraulic fluids and heat exchanges, and as a solvent, on mice, rats, and rabbits was examined by Price et al. (1985) and Tyl et al. (1993). Several variables were measured on the pregnant dam during the experiments including maternal weight change, water consumption, liver weight, and kidney weight. In addition, several measurements were made on each litter including number of live fetuses, fetal malformations, and fetal body weight. Table 5.6 shows the summary statistic for the fetal weight.

Using the fetal weight data on CD-1 mice, Dempster et al. (1984) applied a nested linear model to analyze the data. Note that the weights are nested within litters. The experiment consisted of control and three non-zero doses of 750, 1500, and 3000 mg/kg per day and a total of 94 dams were exposed during the gestational period. The average fetal weights clearly show a dose-response pattern and the average weight decreases monotonically as the dose increases. Fitzmaurice et al. (2011) note that although the weight is not a linear function of dose, a square root transformation of the dose makes is approximately linear; i.e. weight is a linear function of $\sqrt{\text{dose}}$. They use the nested linear model

$$Y_{ij} = \beta_0 + \beta_1 d_j + b_j + \epsilon_{ij}$$

where Y_{ij} is the fetal weight of the ith live fetus in the jth litter and d_j the square root transformed dosage scaled by 750. The term b_j denotes the random effect which is assumed to vary independently across all litters having a normal distribution with mean 0 and standard deviation σ_2. The error term ϵ_{ij} is assumed to have a normal distribution with a mean of 0 and standard deviation of σ_1 and varies independently across fetuses within litters. By fitting the above model, it is found that there is a significant dose effect with an estimated common intralitter correlation of 0.57.

Ryan et al. (1991) discussed the relationship between fetal weight and malformation in developmental toxicity studies. Using data from ten developmental toxicity studies conducted for the National Toxicology Program (NTP), the authors show that there is a clear pattern and

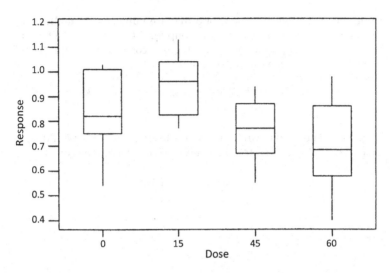

FIGURE 5.2
Boxplots of average fetal weights.

malformed fetuses at term tended to have lower weight than normal
non-malformed fetuses. The pattern demonstrates the potential value in
simultaneous modeling of joint effect of exposure on malformation and
weight. Therefore considering a single continuous outcome in isolation
and not in conjunction with other outcomes is probably not appropriate. We will discuss the joint modeling of multiple responses later in
this chapter. In fact, Razzaghi and Loomis (2001) warn that considering
the fetal weight alone in a single replicate of an experiment can lead to
erroneous conclusions. Data presented in Table 5.6 give the average fetal
weight per litter for a single replicate for the C57BL/6 mice in the experiment described in Example 5.1 that was conducted by the NCTR to assess
the developmental toxicity of 2,4,5-T (see Holson et al., 1992). Figure 5.2
depicts the boxplot of the summary of the data. The central box spans the
quartiles, the line in the box marks the median. From the boxplot, one can
observe the "umbrella"-type dose-response relationship, indicating that
the chemical may have low-dose beneficial effects and that small doses of
2,4,5-T to the pregnant dam may actually increase the birth weight of the
offspring. This assertion may be verified by using a formal testing procedure from Mack and Wolf (1981) where a nonparametric testing procedure is proposed for detecting the so-called umbrella dose-response
effect. Consider testing the null hypothesis $H_0 : \mu_0 = \mu_1 = \cdots = \mu_g$, i.e. that
there is no dose effect and that the average fetal weight is the same for
all dose groups and equivalent to the unexposed control animals against
the alternative that there is an experimental dosage level D such that the
group means are increasing up to and including that dosage level and
then decline after that dosage level. Thus, the alternative hypothesis can
be expressed as $H_1 : \mu_0 \leq \mu_1 \leq \cdots \leq \mu_D \geq \mu_{D+1} \geq \cdots \geq \mu_g$ which represents
the "umbrella" effect. Note that in the alternative hypothesis, some of
the inequalities may be strict. Let U_{kl} be the number of experimental

group means in the *l*th dose group that are numerically smaller than the experimental means in the *k*th group for $l,k = 0,1,...,g$. Define the indicator χ_i as being equal to 1 if the *i*th group corresponds to the "peak" estimated by the sample and 0 otherwise

$$\begin{cases} \chi_i = 1 & \text{if } i\text{th group corresponds to the peak} \\ 0 & \text{otherwise} \end{cases}.$$

Then, Mack and Wolf (1981) show that the testing procedure would reject the null hypothesis of no treatment effect H_0 for large values of

$$A_{\hat{D}} = \sum_{i=0}^{g} \frac{\chi_i(A_i - \mu_0(A_i))}{\sigma_0(A_i)}$$

where

$$A_i = \sum_{1 \le l < k \le i} U_{lk} + \sum_{i \le l < k \le g} U_{kl}$$

and $\mu_0(A_i)$ and $\sigma_0(A_i)$ are respectively the mean and variance of A_i under the null hypothesis H_0. Mack and Wolf (1981) show that $\mu_0(A_i)$ and $\sigma_0(A_i)$ are respectively given by

$$\mu_0(A_i) = \frac{1}{4}\left\{ N_1^2 + N_2^2 - \sum_{t=0}^{g} m_t^2 - m_i^2 \right\}$$

and

$$\sigma_0^2(A_i) = \frac{1}{72}\left\{ 2\left(N_1^3 + N_2^3\right) + 3\left(N_1^2 + N_2^2\right) - \sum_{t=0}^{g} m_t^2\left(2m_t + 3\right) \right.$$
$$\left. - m_i^2\left(2m_i + 3\right) - 12m_i N_1 N_2 - 12m_i^2 N \right\}$$

where, as before, m_i is the number of pregnant dams in the *i*th dose group for $i = 0, 1, ... g$ and $N_1 = \sum_{t=0}^{i} m_t$, $N_2 = \sum_{t=i}^{g} m_t$, and $N = \sum_{t=0}^{g} m_t$. The test is asymptotically normal and can also be used to detect the existence of an inverted umbrella or a U-shaped dose-response curve. One needs to only interchange U_{lk} and U_{kl} in the expression for A_i, leaving all else unchanged. Mack and Wolf (1981) further provide equal sample size critical values for the proposed test using simulation. If the sample sizes are not the same, they assert that based on their Monte Carlo simulation, the large sample approximation is sufficiently accurate to allow one to safely use $m_0 = m_1 = ... m_g = 10$ critical values to approximate the corresponding critical values for any sample size configuration for which N/g is at least 10. Applying this test to the C57BL/6 mice data from Table 5.4, we have

$$\chi_1 = 1, \chi_0 = \chi_2 = \chi_3 = 0$$

TABLE 5.4

Developmental Effect of Diethylhexyl Phthalate (DEHP) in Mice

Dose mg/kg/Body Weight (BW)	0	44	91	191	292
Proportion affected		0.19	0.12	0.25	0.72	0.98

and

$$A_1 = U_{01} + U_{21} + U_{31} + U_{32} = 593$$

from which we have

$$\mu_0(A_1) = 393, \quad \sigma_0(A_1) = 56.4624$$

and thus

$$A_{\hat{1}} = 3.542.$$

Since the average sample size per dose group exceeds 10, we compare this value with critical values given in table 3 of Mack and Wolfe (1981). We find that the test is highly significant ($p \leq .01$) and data indicate an umbrella-type dose-response curve with the peak estimated to be at the dose of 15 mg/kg body weight. Note that the test will necessarily estimate the occurrence of the peak at an experimental dose level though in theory this may not be the case. One can therefore conclude that the estimated peak is the closest experimental dose to the level of exposure corresponding to the peak. Hence, the data suggest that small doses of 2,4,5-T administered to pregnant mice during the gestation period may actually increase the birth weight of the developing fetus. This concept of a chemical having beneficial effects at low doses, becoming toxic at high doses, and hence leading to a nonlinear dose-response relationship is known as "hormesis" and for a long time has attracted the attention of many scientists, for example, Calabrese (2014). Thus, we see here that the test leads to the conclusion that 2,4,5-T may have a hermetic developmental effect. Ryan et al. (1991) show by considering the result of ten developmental toxicity experiments conducted by the National Toxicology Program that there is clear evidence that malformed fetuses tend to have lighter weight at term than non-malformed fetuses. Their findings suggest that fetal weight provides a sensitive and important outcome for assessing developmental toxicity, especially in rodents. The correlation between fetal weight and other endpoints in developmental toxicity experiments was also considered by Chen and Gaylor (1992).

The above analysis may lead to the conclusion that 2,4,5-T has a hormetic effect and that small doses of the chemical can lead to a lower proportion of malformed fetuses in C57BL/6 mice. However, such a conclusion is erroneous and can have catastrophic consequences. Table 5.5 presents the means and

TABLE 5.5

Developmental Toxicity of Diethylhexyl Phthalate (DEHP)

	Dose = 0			Dose = 0.091			Dose = 0.292			Dose = 0.044			Dose = 0.191	
Litter Number	Number Live	Number Affected	Litter Number	Number Live	Number Affected	Litter Number	Number Live	Number Affected	Litter Number	Number Live	Number Affected	Litter Number	Number Live	Number Affected
1	8	5	1	6	3	1	9	9	1	4	1	1	1	1
2	9	2	2	8	0	2	10	10	2	8	3	2	2	2
3	10	1	3	8	2	3	10	10	3	9	0	3	2	2
4	10	1	4	9	3	4	11	11	4	9	2	4	4	4
5	11	0	5	11	1	5	11	11	5	10	1	5	6	6
6	11	3	6	11	2	6	11	11	6	11	0	6	10	3
7	12	0	7	11	4	7	11	11	7	11	1	7	10	10
8	12	0	8	11	5	8	11	11	8	12	1	8	11	2
9	12	2	9	11	6	9	12	12	9	12	1	9	11	4
10	12	4	10	12	0	10	12	12	10	12	2	10	11	6
11	12	5	11	12	1	11	12	12	11	13	0	11	11	9
12	13	0	12	12	2	12	12	12	12	13	0	12	12	10
13	13	0	13	12	3	13	12	12	13	13	1	13	12	12
14	13	0	14	13	1	14	12	12	14	13	1	14	12	12
15	13	1	15	13	4	15	13	12	15	13	1	15	13	5
16	13	2	16	14	0	16	13	13	16	13	1	16	13	12
17	13	3	17	14	2	17	13	13	17	13	3	17	14	2
18	13	4	18	14	2	18	13	13	18	13	4	18	14	6
19	14	0	19	14	4	19	13	13	19	14	1	19	14	10
20	14	1	20	14	5	20	13	13	20	14	1	20	14	11
21	14	3	21	14	8	21	14	14	21	14	2	21	14	11
22	14	5	22	14	9	22	14	14	22	15	1	22	15	3

(Continued)

TABLE 5.5 (CONTINUED)

Developmental Toxicity of Diethylhexyl Phthalate (DEHP)

Dose = 0			Dose = 0.091			Dose = 0.292			Dose = 0.044			Dose = 0.191		
Litter Number	Number Live	Number Affected	Litter Number	Number Live	Number Affected	Litter Number	Number Live	Number Affected	Litter Number	Number Live	Number Affected	Litter Number	Number Live	Number Affected
23	15	0	23	15	2	23	16	16	23	15	1	23	15	15
24	16	0	24	15	2	24	17	17	24	15	2	24	18	17
25	16	1	25	15	3				25	15	4	25	19	19
26	16	1	26	16	6				26	16	2			
27	16	9												
28	16	10												
29	17	1												
30	18	11												

Data from Tyl et al. (1993).

standard deviations of the average fetal body weights for the viable fetuses together with the number of litters at different dose levels for the C57BL/6 mice in the entire experiment. In a discussion regarding the existence of a threshold dose in teratogenesis, Gaylor et al. (1988) argue that if a chemical produces a developmental defect by augmenting or accelerating an already existing mechanism that produces spontaneous defects, then no population threshold dose can exist and any dose level above the control can result in adverse effects. This means that a threshold dose or even a hormetic effect can exist only if it can be proved that the mechanism of toxicity at the background dose is different from those at higher doses. Gaylor et al. (1988) further argue that if a threshold dose exists, then its value is different for each individual in a population and changes from day to day depending on other environmental insults. This implies that the threshold dose for a population resulting in absolutely no risk (or even some beneficial effects) must take into account the most sensitive environmental conditions for the most sensitive individual. It is pointed out that the 2,4,5-T study by Holsen et al. (1992) described above failed to demonstrate threshold doses in any of the strains. Actually, Holson et al. (1992) used linear regression to investigate the relationship between the mean fetal body weight and the dose of 2,4,5-T based on data from every strain. They show that all the regression lines have negative slopes and that the dose response decreases with fetal weight. In fact Holson et al. (1992) further demonstrate that 2,4,5-T has an adverse effect on fetuses of every strain of mice studied in the experiment and that developmental toxicity was mediated through similar mechanisms since there was no significant difference between the negative slopes of the regression dose-response lines for most of the strains including C57BL/6. Moreover, they found that there was substantial variation among replicates within the same strain. Since disparity among individual replicates can result in different conclusions, multireplicated experiments are recommended for all developmental toxicants.

5.8 Modeling Count

It is not uncommon to consider only a count such as the number of live fetuses as the outcome of the developmental toxicity study. Let n_{ij} be the observed number of live fetuses in the jth litter of the ith dose group. Then a natural place to start is to model the counts as a Poisson distribution, i.e.

$$P(n_{ij} \mid \mu_i) = \frac{\mu_i^{n_{ij}} e^{-\mu_i}}{n_{ij}!} \quad n_{ij} = 0, 1, \ldots$$

where $\mu_i > 0$ is the mean number of live fetuses among all litters in the ith dose group. Now, the mean function μ_i can be modeled appropriately as a function of dose. Most commonly, if we assume

$$\mu_i = \exp(\beta_0 + \beta_1 d_i) \quad i = 0,1,\ldots,g$$

which is the log-linear link, then the familiar Poisson regression model is resulted. This modeling approach, however, is not appropriate since the Poisson model forces the equality of mean and variance. As discussed before, data from developmental toxicity experiments exhibit overdispersion and ignoring the extra Poisson variation can lead to serious errors. Although there are many approaches to incorporate the overdispersion in count modeling, perhaps the most common approach is to use a gamma-Poisson mixture. If we assume a gamma distribution with mean 1 and variance of $(\varphi_i)^{-1}$ for the mean function μ_i that is,

$$g(\mu_i) = \frac{1}{\varphi_i^{\varphi_i} \Gamma(\varphi_i)} \mu_i^{\varphi_i - 1} e^{-\mu_i / \varphi_i}$$

then the unconditional distribution of n_{ij} is the gamma-Poisson mixture given by

$$P(n_{ij}) = \frac{\Gamma\left(n_{ij} + \dfrac{1}{\varphi_i}\right)}{\Gamma\left(n_{ij} + 1\right)\Gamma\left(\dfrac{1}{\varphi_i}\right)} \left[\frac{\varphi_i \mu_i}{1 + \varphi_i \mu_i}\right]^{n_{ij}} \cdot \left[\frac{1}{1 + \varphi_i \mu_i}\right]^{\frac{1}{\varphi_i}} \quad \varphi_i > 0$$

which belongs to the negative binomial family of distributions. For this model,

$$E(n_{ij}) = \mu_i \tag{5.14}$$

and

$$V(n_{ij}) = \mu_i(1 + \varphi_i \mu_i) \tag{5.15}$$

As in the case of the beta-binomial model for binary data discussed earlier in this chapter, one problem with the above modeling procedure is that the assumption of $\varphi_i > 0$ is restrictive. One remedy to overcome this problem is to apply the GEE methodology with a mean and variance function defined by (5.14) and (5.15) above.

Example 5.5

The developmental toxicity effects of DEHP were described in Tyl et al. (1983). Chen and Kodell (1989) used the pup-specific data from that experiment to fit a beta-binomial distribution along with a Weibull dose-response model. Table 5.3 gives the number of live fetuses as well as the number affected per litter at each dose level. Tables 5.4 and 5.5 display the R outputs for Poisson regression and Poisson-gamma regression of

TABLE 5.6

Weight Distribution of the Fetal Weight by Dose in Mice (Ethylene Glycol)

Dose	No of Litters	No of Fetuses	Mean Weight	SD of Weights
0	25	297	0.972	0.098
750	24	276	0.877	0.104
1500	22	229	0.764	0.107
3000	23	226	0.704	0.124

the DEHP data using the number of live fetuses as the response variable and the dosage level as the only explanatory variable. We can clearly see that using only the number of live fetuses would not provide sufficient information for analyzing the data. In fact, from Table 5.6 where the mean and variance of the number of live fetuses is calculated for each dose group, we can see that even the overdispersed nature of the data is not evident from the number of live fetuses. Count data modeling has not been well investigated for developmental experiments.

5.9 Statistical Models for Trinomial Outcomes

So far in this chapter, we have considered models for a single binary outcome, that is when fetuses are classified as affected or unaffected, and were "affected" outcome included both dead/resorbed as well as malformed fetuses. In reality, the two outcomes "dead/resorbed" and "malformed" should be considered separately because of the hierarchical structure of these outcomes. More specifically, in order for a fetus to become viable, it has to survive the dead/resorption stage. Therefore the outcome of "malformed" is conditionally dependent on a fetus being viable. Assuming that these outcomes are independent may lead to erroneous results. It is therefore more appropriate to consider the trinomial structure of the outcomes. Accordingly, let $Y_{ij} = (Y_{ij1}, Y_{ij2}, Y_{ij3})$ be the observed vector of the jth dam in the ith dose group, where the components of the vector Y_{ij} represent the number of dead/resorbed, malformed, and normal fetuses respectively. Clearly,

$$Y_{ij1} + Y_{ij2} + Y_{ij3} = n_{ij} \quad j = 1,\dots,m_i, i = 1,\dots,g.$$

Following Catalano and Ryan (1992), let $\theta_1(d_i)$ be the probability that at exposure level of d_i; $i = 0, 1, \dots, g$ an implanted embryo dies or is resorbed during the gestation. Thus, the probability $P_1(d_i)$ associated with Y_{ij1} is given by

$$P_1(d_i) = \theta_1(d_i).$$

Given that a fetus survives death/resorption, let the conditional probability that it becomes malformed be $\theta_2(d_i)$. Then the probability associated with Y_{ij2} is given by

$$P_2(d_i) = [1 - \theta_1(d_i)]\theta_2(d_i).$$

Finally, the probability that a fetus is born normal, associated with Y_{ij3}, can be expressed as

$$P_3(d_i) = [1 - \theta_1(d_i)][1 - \theta_2(d_i)].$$

One of the earliest attempts of simultaneously modeling trinomial responses from developmental toxicity experiments is by Chen et al. (1991) who considered a Dirichlet-trinomial model as a natural extension of the beta-binomial model. Assume a trinomial distribution for the response vector Y_{ij} that is

$$P\left(y_{ij1}, y_{ij2}, y_{ij3} \mid p_{ij1}, p_{ij2}, p_{ij3}\right) = \binom{n_{ij}}{y_{ij1}, y_{ij2}} p_{ij1}^{y_{ij1}} p_{ij2}^{y_{ij2}} p_{ij3}^{y_{ij3}}$$

where $p_{ij1}, p_{ij2}, p_{ij3}$ are respectively the probabilities that the jth litter of the ith dose group is dead/resorbed, malformed, or normal with $p_{ij1} + p_{ij2} + p_{ij3} = 1$ and

$$\binom{n_{ij}}{y_{ij1}, y_{ij2}} = \frac{n_{ij}!}{y_{ij1}! y_{ij2}! \left(n_{ij} - y_{ij1} - y_{ij2}\right)!}.$$

Also, if we express the joint distribution of $p_{ij1}, p_{ij2}, p_{ij3}$ as the Dirichlet distribution given by

$$P\left(p_{ij1}, p_{ij2}, p_{ij3}\right) = \frac{\Gamma\left(\alpha_i + \beta_i + \gamma_i\right)}{\Gamma(\alpha_i)\Gamma(\beta_i)\Gamma(\gamma_i)} p_{ij1}^{\alpha_i - 1} p_{ij2}^{\beta_i - 1} p_{ij3}^{\gamma_i - 1}$$

$$\alpha_i, \beta_i, \gamma_i > 0$$

then the unconditional distribution of Y_{ij} is given by

$$P\left(y_{ij1}, y_{ij2}, y_{ij3}\right) = \frac{n_{ij}! \Gamma\left(\alpha_i + \beta_i + \gamma_i\right) \Gamma\left(y_{ij1} + \alpha_i\right) \Gamma\left(y_{ij2} + \beta_i\right) \Gamma\left(y_{ij3} + \gamma_i\right)}{y_{ij1}! y_{ij2}! y_{ij3}! \Gamma\left(n_{ij} + \alpha_i + \beta_i + \gamma_i\right) \Gamma(\alpha_i)\Gamma(\beta_i)\Gamma(\gamma_i)} \tag{5.16}$$

which clearly is a generalization of the beta-binomial distribution, called the Dirichlet-trinomial distribution. Under the Dirichlet-trinomial model, we have $E(Y_{ij1}) = n_{ij}\mu_{i1}$ and $E\left(Y_{ij2} \mid y_{ij1}\right) = \left(n_{ij} - y_{ij1}\right)\mu_{i2}$ where $\mu_{i1} = \alpha_i\theta_i$ and

$\mu_{i2} = \beta_i \theta_i (1 - \alpha_i \theta_i)$ with $\theta_i = (\alpha_i + \beta_i + \gamma_i)^{-1}$. The intralitter correlation for the Dirichlet-trinomial model is $\rho_i = \dfrac{\theta_i}{1 + \theta_i}$. The parameters μ_{i1} and μ_{i2} can be interpreted as the means p_{ij1} and p_{ij2} respectively. Since in the above model y_{ij1}, y_{ij2}, and y_{ij3} are mutually negatively correlated, the Dirichlet-trinomial model assumes that the correlations between the number of deaths/resorptions, malformations, and normal fetuses are negative. Several properties of the Dirichlet-trinomial distribution in modeling the multiple outcomes from developmental toxicity experiments are discussed by Chen et al. (1991). Specifically, the authors show (see Exercise 5 at the end of this chapter) that the marginal distribution of y_{ij1}, the number of dead/resorbed fetuses is a beta-binomial distribution with parameters n_{ij}, α_i, and β_i. The marginal distribution of Y_{ij3}, the number of normal fetuses is a beta-binomial distribution with parameters n_{ij}, γ_i, and $\alpha_i + \beta_i$. Also, the conditional distribution of Y_{ij2}, the number of malformed fetuses given Y_{ij1}, the number of dead/resorbed fetuses is a beta-binomial distribution with parameters $(n_{ij} - y_{ij1})$, β_i, and γ_i. Further, the Dirichlet-trinomial distribution can be expressed as the product of two beta-binomial models and thus it is a special case of the more general class of distributions called the double beta-normal

$$P\left(y_{ij1}, y_{ij2}, y_{ij3}\right) = \frac{n_{ij}!\,\Gamma\left(\alpha_i + \delta_i\right)\Gamma\left(y_{ij1} + \alpha_i\right)\Gamma\left(n_{ij} - y_{ij1} + \delta_i\right)}{y_{ij1}!\left(n_{ij} - y_{ij1}\right)\Gamma\left(n_{ij} + \alpha_i + \gamma_i\right)\Gamma(\alpha_i)\Gamma(\delta_i)} \cdot$$

$$\frac{(n_{ij} - y_{ij1})!\,\Gamma\left(\beta_i + \gamma_i\right)\Gamma\left(y_{ij2} + \beta_i\right)\Gamma\left(n_{ij} - y_{ij1} - y_{ij2} + \gamma_i\right)}{y_{ij2}!\left(n_{ij} - y_{ij1} - y_{ij2}\right)!\,\Gamma\left(n_{ij} - y_{ij1} + \beta_i + \gamma_i\right)\Gamma(\beta_i)\Gamma(\gamma_i)} \qquad (5.17)$$

with constraint $\delta_i = \beta_i + \gamma_i$ (see Exercise 6 at the end of this chapter). For the purpose of risk assessment with trinomial responses, the first question is how to define the overall risk. Catalano et al. (1993) suggest that the overall risk can be defined as the probability of being affected, i.e. being either dead/resorbed or malformed. Thus,

$$P(d) = 1 - P_3(d) = 1 - [1 - \theta_1(d)][1 - \theta_2(d)].$$

Now, in the Dirichlet-trinomial model given in (5.16) we first derive a reparameterization by substituting

$$\alpha_i = \frac{\mu_{i1}}{\theta_i}$$

$$\beta_i = \frac{\mu_{i2}}{\theta_i(1 - \mu_{i1})}$$

$$\gamma_i = \frac{(1-\mu_{i1})^2 - \mu_{i2}}{\theta_i(1-\mu_{i1})}$$

The mean functions μ_{i1} and μ_{i2} are then replaced by suitable dose-response functions and model parameters are estimated.

Example 5.6

The Dirichlet-trinomial model and the double beta-binomial model were used by Chen and Li (1994) for simultaneous modeling of trinomial responses in developmental toxicity experiments. Using logistic dose-response models for both μ_{i1} and μ_{i2},

$$\mathrm{logit}(\mu_{ij}) = \log\frac{\mu_{ij}}{1-\mu_{ij}} = \gamma_{0j} + \gamma_{1j}d_i \quad j = 1,2$$

along with the quasi-likelihood approach, the parameter estimates are derived. Note that both the Dirichlet-trinomial model and the double beta-binomial models assume that proportion of death/resorbed and the proportion of malformed are independent. The Dirichlet-trinomial model uses a common intralitter correlation for both endpoints while there are two intralitter correlation parameters in the double beta-binomial model, one for each endpoint. Although no risk assessment was performed, two data sets were utilized to illustrate the application of the quasi-likelihood approach.

Joint modeling of multiple multivariate binary outcomes from developmental toxicity studies was also considered by other authors. For example Mohlenbergs and Ryan (1999) defined a multivariate exponential family model that in addition to multiple outcomes, allowed for clustering. The model accounted for main effects and first order interactions. Hunt et al. (2008) propose threshold models and spline models as a dose response along with a likelihood approach for estimating parameters. The methodology was illustrated using the DEHP and diethylene glycol dimethyl ether (DYME) (Price et al., 1987) data sets. No specific model has been universally accepted as the optimal.

5.10 Statistical Models for Multiple Binary and Continuous Outcomes

It should be realized that consideration of a single binary or continuous outcome, or even models that consider only multiple binary outcomes do not truly characterize developmental toxicity experiments. Using the results of ten studies from the NTP, Ryan et al. (1991) found a direct correlation between the incidence of malformation and fetal weight. Consistently, normal fetuses had higher weight and malformed fetuses had lower weight. This suggests that, in reality, joint modeling of fetal weight and malformation would better characterize the outcomes of developmental toxicity studies. The difficulty, however, is that the fetal weight is a continuous outcome while the incidence of malformation is a binary and a discrete variable, and therefore we need models for joint consideration of discrete and continuous outcomes. Although some authors had discussed joint modeling of discrete and continuous outcomes, they were not applicable to developmental studies since the models did not account for the intralitter correlation. Besides the models were developed with a focus on deriving a single p-value for the experimental results rather than exploring the relationship between outcomes. Perhaps the first attempt to jointly model fetal weight and malformation is by Catalano and Ryan (1992) who define an unobserved latent variable model for the discrete outcome. Accordingly let the random variable Y_{ij} be the value of the latent variable for the jth fetus in the ith litter. That is, suppose that there is a threshold, which without the loss of generality can be assumed to be 0. For otherwise a shift in the data would make the adjustment above which the fetus becomes malformed. Thus if we define Y_{ij}^* as the binary malformation indicator, we have

$$Y_{ij}^* = \begin{cases} 1 & Y_{ij} > 0 \\ 0 & Y_{ij} \leq 0 \end{cases}.$$

Now, assume a linear model for Y_{ij},

$$Y_{ij} = \beta_0 + \beta_1 d_i + \epsilon_{ij} \quad j = 1,\ldots,m_i; \ i = 0,1,\ldots,g$$

where d_i is the dosage administered to the ith pregnant animal. Note that while Y_{ij}^* is observed from the experiment, Y_{ij} is not observable. Let W_{ij} be the fetal weight of the jth fetus in the ith litter and consider a linear model for W_{ij},

$$W_{ij} = \alpha_0 + \alpha_1 d_i + \delta_{ij} \quad j = 1,\ldots,m_i; \ i = 0,1,\ldots,g.$$

The simplest approach for joint modeling of malformation and fetal weight is to use the normal linear regression model and assume a bivariate normal distribution for δ_{ij} and ϵ_{ij}

$$\begin{pmatrix} \delta_{ij} \\ \epsilon_{ij} \end{pmatrix} \sim N\left(\begin{pmatrix} 0 \\ 0 \end{pmatrix}, \begin{pmatrix} \sigma_\delta^2 & \rho\sigma_\epsilon\sigma_\delta \\ \rho\sigma_\epsilon\sigma_\delta & \sigma_\epsilon^2 \end{pmatrix} \right).$$

This approach assumes a constant correlation ρ between W_{ij} and Y_{ij}. However, it assumes that littermates are independent and ignores the intralitter correlation. Note that by the property of the bivariate normal distribution we have

$$P\left(Y_{ij}^{*}\right) = \Phi\left(\frac{\beta_0 + \beta_1 d_i}{\sigma_\epsilon}\right)$$

which is the probit model.

Now, the joint distribution of Y_{ij} and Y_{ij}^{*} can be expressed in terms of the product of marginal and conditional distributions as

$$f(w, y^{*}) = f(w) \cdot g(y^{*} \mid w) \tag{5.18}$$

where $f(w)$ is the marginal distribution of W_{ij} and $g(y \mid w)$ is the conditional distribution of Y_{ij}^{*}. Further, by the standard properties of the bivariate normal distribution we have that the conditional distribution of Y_{ij} given W_{ij} is normal with mean

$$\mu = \beta_0 + \beta_1 d_i + \left(\frac{\sigma_\epsilon}{\sigma_\delta}\right) \rho \, e_{ij} \tag{5.19}$$

and variance

$$\sigma_\epsilon^2 \left(1 - \rho^2\right)$$

where e_{ij} is the residual

$$e_{ij} = W_{ij} - \left(\alpha_0 + \alpha_1 d_i\right)$$

from the weight model. Further, the conditional distribution of Y_{ij}^{*} given W_{ij} is the probit model given by

$$P(Y_{ij}^{*} = 1 \mid W_{ij}) = \Phi\left(\frac{\mu}{\sigma_\epsilon \sqrt{1 - \rho^2}}\right)$$

where, as noted by Catalano and Ryan (1992), we see that if there is perfect correlation between fetal weight and malformation, the above conditional distribution becomes degenerate. Substituting for μ from (5.19), we have

$$P\left(Y_{ij}^{*} = 1 \mid W_{ij}\right) = \Phi\left(\beta_0^{*} + \beta_1^{*} d_i + \beta_2^{*} e_{ij}\right) \tag{5.20}$$

where

$$\beta_0^{*} = \frac{\beta_0}{\sigma_\epsilon \sqrt{1 - \rho^2}}, \beta_1^{*} = \frac{\beta_1}{\sigma_\epsilon \sqrt{1 - \rho^2}}, \beta_2^{*} = \frac{\rho}{\sigma_\delta \sqrt{1 - \rho^2}}.$$

Equation (5.20) defines a generalized linear regression model for malformation with the probit link in which the residual from the weight model is a covariate in addition to the administered dose. Thus, in practice, the parameters β_0^*, β_1^*, and β_2^* can be estimated using the usual regression modeling. It is interesting to note in (5.20) that when $\rho = 0$, the equation collapses into the unconditional probit model.

Since the above modeling structure ignores the important issue of litter effect, Catalano and Ryan (1992) consider an extension and derive a model similar to (5.20) with a slightly different covariate structure. In the extended model, both individual and litter averages of residuals are used as covariates

$$P\left(Y_{ij}^* = 1 \mid W_{ij}\right) = \Phi\left(\beta_0^* + \beta_1^* d_i + \beta_2^* \overline{e_{ij}} + \beta_3^* \left(e_{ij} - \overline{e_{ij}}\right)\right) \qquad (5.21)$$

Thus, in practice, we have a pair of regression models. The first models the fetal weight as a linear function of the administered dose, and the second regresses malformation outcome conditional on fetal weight outcome as a function of dose using the residuals from the first model as covariates. The regression coefficients in the second model are directly related to the variance and correlation structure of the defined latent variable. To illustrate the modeling procedure, a data set from a developmental toxicity experiment in mice conducted through the NTP is used. The experiment is the study of the developmental effects of ethylene glycol (EG) described by Price et al. (1985). Table 5.7 provides a summary of the data over all litters (no litter effect), and the information is retrieved from table 5 of Price et al. (1985). See also Table 5.8. The experiment consisted of exposing pregnant CD-1 mice during gestation days 6 through 15. Several other variables were also measured.

To apply the two-step modeling procedure proposed by Catalanao and Ryan (1992) in the context of risk assessment with malformation and fetal weight as outcomes, Catalano et al. (1993) define an affected fetus as a fetus which is either dead/resorbed, or has a malformation, or has a below-normal fetal weight. Thus if $\theta_1(d)$ and $\theta_2(d)$ respectively denote the probability of death/resorption and probability of malformation and/or low fetal weight, the probability that a fetus is abnormal is given by

$$P(d) = 1 - \left[1 - \theta_1(d)\right]\left[1 - \theta_2(d)\right] \qquad (5.22)$$

Note that this is similar to how we defined the probability of being abnormal when we considered the joint modeling of trinomial responses except that now $\theta_2(d)$ is defined differently to include low fetal weight as well. For parametric models, Catalano et al. (1993) suggest using a modified probit model for the binary variable of death/resorption and a regression model for a power of dose with litter size as covariate. After fitting the regression model, according to the modeling procedure, individual and average residuals are calculated and the conditional distribution of the malformation

TABLE 5.7

Average Fetal Weight per Litter (Grams) in a Single Replicate

Litter	Dose			
	0	15	45	60
1	0.54	0.65	0.73	0.57
2	0.69	0.79	0.80	0.40
3	0.78	1.06	0.87	0.56
4	0.88	1.03	0.67	0.60
5	0.83	0.96	0.77	0.40
6	1.03	0.87	0.78	0.63
7	0.76	0.96	0.71	0.98
8	1.01	0.77	0.55	0.71
9	1.01	0.85	0.62	0.62
10	0.82	0.80	0.68	0.88
11	0.75	1.04	0.90	0.93
12		1.13	0.67	0.75
13		1.04	0.90	0.95
14			0.83	0.77
15			0.94	0.66
16				0.80
Average	0.83 g	0.93	0.76	0.70

TABLE 5.8

Average Fetal Weight for All Litters (Grams)

Dose	Number of Litters	Average Fetal Weight	Standard Deviation
0	71	82.34	15
15	13	93.57	12
30	54	74.15	16
45	74	71.19	13
60	69	67.58	14
75	37	59.91	16
90	19	56.27	10

indicator variable is modeled using another modified probit model. Finally, to model the probability of low fetal weight, Catalano et al. (1993) suggest choosing a threshold response such as three standard deviations below the mean and call fetuses, with fetal weight lower than this threshold, abnormal. Thus another modified probit model is used for parameterization of this probability. Combining all models and using (5.21), the following dose-response function is derived for risk assessment

$$P(d) = 1 - \left[1 - \Phi\left(\gamma_0 + \gamma_1 d^{\gamma_2}\right)\right]\left[1 - \Phi\left(\beta_0 + \beta_1 d^{\beta_2}\right)\right]\left[1 - \Phi\left(\frac{w_0 - \left(\alpha_0 + \alpha_1 d^{\alpha_2}\right)}{\sigma_w}\right)\right]$$

where w_0 is the fetal weight threshold below which a fetus is classified as abnormal and σ_w is the standard deviation of fetal weight in control animals. By applying the methodology to an experimental data set regarding the developmental effects of exposure to diethylene glycol dimethyl ether (Price et al., 1987) the authors show that there is a strong correlation between different fetal outcomes and that a combined analysis considering all possible outcomes including death/resorption, incidence of malformation, and fetal weight provides a "useful and sensitive summary" of the overall toxicity effects of the chemical. Furthermore, the model provides a better understanding of the underlying relationship between different outcomes.

Fitzmaurice and Laird (1995) argue that in the model proposed by Catalano and Ryan (1992), the regression parameters for both the binary malformation status and conditional fetal weight outcomes have no specific marginal interpretation. This is due to the fact that the link function relating the conditional mean of the binary response to the covariates is nonlinear. They propose a model to describe the marginal distribution of y_{ij} given the covariates as a logistic model given by

$$f(y) = \frac{e^{\theta_i y}}{1 + e^{\theta_i y}} \tag{5.23}$$

where it is assumed that θ_i is a linear function of a set of covariates S_i predicting y_{ij} and is related to the mean response $\mu_i = \mu_i(\beta_1) = E(Y_{ij}) = P(Y_{ij} = 1)$ through a logistic link, that is

$$\theta_i = \log \frac{\mu_i}{1 - \mu_i} = S_i \beta_1.$$

Writing the joint distribution of W_{ij} and Y_{ij} as

$$f(w, y) = f(y) \cdot g(w|y) \tag{5.24}$$

and assuming a normal for the conditional distribution of W_{ij},

$$g(w|y) = \frac{1}{2\pi\sigma^2} \exp\left\{ -\frac{1}{2\sigma^2} \left[w - \mu_w - \alpha(y - \mu_i) \right]^2 \right\} \tag{5.25}$$

where μ_w is a linear function of another set of covariates T_i predicting the weight, i.e.

$$\mu_w = T_i \beta_2$$

and α is a parameter for regression of W_{ij} on Y_{ij}, it becomes apparent that the weight variable W_{ij} has a conditional mean that depends on the binary malformation variable Y_{ij}, inducing the correlation between the two variables. This modeling structure can be extended to allow for clustering by using the

GEE methodology. Fitzmaurice and Laird (1995) illustrate their methodology by using the experimental data on the developmental effects of exposure to diethylene glycol dimethyl ether (Price et al., 1985).

Both, Catalano and Ryan (1992) and Fitzmaurice and Laird (1995) rely on factorization of the joint distribution of the binary variable for malformation and continuous variable for fetal weight as the product of a marginal distribution and a conditional distribution. In fact, a couple of other models were developed based on this conditioning structure; see Chen (1993) and Ahn and Chen (1997). But, as noted in Regan and Catalano (1999), this conditioning is largely because of statistical convenience and not so much based on biological principles. In addition, conditional models do not provide adequate interpretation of the marginal dose-response relationships, fail to provide an estimate of the direct correlation between fetal weight and malformation status, and are often difficult to apply for risk assessment. Noting these drawbacks, Regan and Catalano (1999) propose a likelihood-based approach that utilizes the joint distribution of a sequence of latent variables induced by binary variables generated by the malformation outcomes in each litter and an extension of the so-called correlated probit model. More specifically, let Y_{jk} be the latent variable corresponding to the malformation status of the kth fetus in the jth litter, i.e. $Y_{jk} > 0$ if the kth pup in the jth litter is malformed and $Y_{jk} \leq 0$ otherwise. Assume that the vector $Y_j = \left(Y_{j1}, \ldots, Y_{m_j} \right)^T$ has a multivariate normal distribution with mean $\mu_y 1_{m_j}$ and covariance matrix $\Sigma_y = \sigma_y^2 \left[\left(1 - \rho_y \right) I_{m_j} + \rho_y J_{m_j} \right]$ where 1_{m_j} is the m_j dimensional vector of ones and I_{m_j} and J_{m_j} respectively represent the $m_j \times m_j$ identity matrix and the matrix of ones. Note that this model assumes equal correlation across all litters. Assuming a multivariate normal distribution for the fetal weight vector $W_j = \left(w_{j1}, \ldots, w_{m_j} \right)^T$, the joint distribution of the $2m_j$ dimensional random vector $\left(Y_j, W_j \right)$ is expressed as a multivariate normal

$$
(2\pi)^{-m_j} \left| \Sigma_{2m_j} \right|^{-\frac{1}{2}} \exp \left\{ -\frac{1}{2} \begin{pmatrix} W_j - \mu_w 1_{m_j} \\ Y_j - \mu_y 1_{m_j} \end{pmatrix}^T \Sigma_{2m_j}^{-1} \begin{pmatrix} W_j - \mu_w 1_{m_j} \\ Y_j - \mu_y 1_{m_j} \end{pmatrix} \right\}
$$

where Σ_{2m_j} is the $2m_j \times 2m_j$ joint covariance matrix of $\left(Y_j, W_j \right)$ given by

$$
\Sigma_{2m_j} = \text{cov}\left(Y_j, W_j \right) = \begin{pmatrix} \sigma_w^2 \left[\left(1 - \rho_w \right) I_{m_j} + \rho_w J_{m_j} \right] & \rho \sigma_w \sigma_y J_{m_j} \\ \rho \sigma_w \sigma_y J_{m_j} & \sigma_y^2 \left[\left(1 - \rho_y \right) I_{m_j} + \rho_y J_{m_j} \right] \end{pmatrix}
$$

and μ_w, σ_w^2, and ρ_w respectively represent the mean, variance, and the common correlation coefficient of components of W_j. Also, ρ represents the correlation coefficient between weight and the latent variable induced by malformation status for each fetus and assumed to be the same for all fetuses within a litter. In this formulation, it is assumed that all outcomes are exchangeable. Regan and Catalano (1999) assume quadratic dose-response models for the mean and the inverse of the coefficient of variation and a linear dose-response model for the variance of the weight variables. They also model all three correlation coefficients linearly in dose and use maximum likelihood to determine the parameter estimates. The methodology is applied to the data from the developmental toxicity of diethylene glycol dimethyl ether by Price et al. (1985) summarized in Tables 5.9 and 5.10. The advantage of this approach in comparison to the conditional modeling approach is that

TABLE 5.9

R Output for Poisson Regression of Diethylhexyl Phthalate (DEHP) Data

```
Call:
glm(formula = resp ~ dose, family = poisson)

Deviance Residuals:
    Min       1Q    Median        3Q       Max
-4.1250   -0.2906   0.1091    0.4550    1.8632

Coefficients:
            Estimate Std. Error z value Pr(>|z|)
(Intercept)  2.54286    0.03692  68.868   <2e-16 ***
dose        -0.30689    0.24012  -1.278    0.201
---
Signif. codes:  0 '***' 0.001 '**' 0.01 '*' 0.05 '.' 0.1 ' ' 1

(Dispersion parameter for Poisson family taken to be 1)

    Null deviance: 115.44  on 130  degrees of freedom
Residual deviance: 113.80  on 129  degrees of freedom
AIC: 682.4

Number of Fisher Scoring iterations: 4
```

TABLE 5.10

R Output for Poisson-Gamma Regression of Diethylhexyl Phthalate (DEHP) Data

```
Call:
glm(formula = resp ~ dose, family = negative.binomial(theta = 1))

Deviance Residuals:
     Min        1Q    Median         3Q        Max
-1.58662   -0.08229   0.02910    0.12230    0.48055

Coefficients:
            Estimate Std. Error t value Pr(>|t|)
(Intercept)  2.54189    0.03189  79.708   <2e-16 ***
dose        -0.29842    0.20358  -1.466    0.145
---
Signif. codes:  0 '***' 0.001 '**' 0.01 '*' 0.05 '.' 0.1 ' ' 1

(Dispersion parameter for Negative Binomial(1) family taken to be
 0.05505775)

    Null deviance: 11.796  on 130  degrees of freedom
Residual deviance: 11.675  on 129  degrees of freedom
AIC: 933.25
```

it allows for estimation of the parameters that characterize the latent variables, which are not directly observable. In addition, since the weight variable and the latent variable induced by malformation status are modeled as linear functions of the dose, they can vary at different dose levels.

There are also some other approaches for joint modeling of discrete and continuous outcomes in developmental toxicity studies. For example, Faes et al. (2004) propose a likelihood-based model using an extension of the Placket-Dale approach to modeling.

5.11 Bayesian Approaches

In the fifteen or so years, Bayesian approaches have found widespread poularity in risk assessment in general ans specifically in developmental toxicity studies. Faes et al. (2006) describe a Bayesian methodology that accounts for the hierarchical structure of the outcomes in developmental toxicology. For an administered dose d, let n_j and u_j be the total number of implants and number of fetal deaths in the jth litter. Then $s_j = n_j - u_j$ is the number of viable fetuses and denote the observations on weight and malformation status of the kth pup by $z_{jk} = (w_{jk}, y_{jk})$ where, as before, $y_{jk} = 1$ if the kth pup has a malformation and 0 otherwise. Given the litter size n_j assume that the distribution of z_{jk} is some pre-specified distribution $F_i(.\,|\,\beta, s_j)$ possibly depending on some covariates such as the dosage level with density $f_i(z_{ij}\,|\,\beta, s_j)$. Suppose that the number of non-viable fetuses u_j is modeled according to a distribution $G(.\,|\,\gamma, n_j)$ with density $g(u_j\,|\,\gamma, n_j)$. This hierarchy lends itself to a Bayesian structure and unknown parameters can be assigned before distribution to derive the posterior models for inferences, given the data. Now, to model $g(u_j\,|\,\gamma, n_j)$, assume a beta-binomial distribution for the number of non-viable fetuses. Using latent variables y_{jk}^* for the malformation status variable y_{jk}, use a normal distribution for modeling the latent variables induced by the malformation variable, that is let

$$\pi_{y_{jk}} = P(Y_{jk} = 1) = P\left(Y_{jk}^* > 0\right) = \Phi\left(\delta_{y_{jk}}\right)$$

where $\delta_{y_{jk}}$ is some parameterized function of predictors such as the administered dosage level and/or the litter size. Next, model the joint distribution of fetal weight and latent variable induced by the malformation variable by a bivariate normal distribution for which the mean and variance of the latent variable are $\delta_{y_{jk}}$ and 1 respectively. Then, using the conditional approach (5.24) Faes et al. (2006) show that the joint distribution of fetal weight and malformation outcome can be expressed as

$$f(w_{jk}, y_{jk}) = f(w_{jk}) \cdot f(y_{jk}\,|\,w_{jk})$$

$$= \phi\left(w_{jk},\,|\,\mu_{w_{jk}}, \sigma_{w_{jk}}^2\right) \cdot \pi_{y|w_{jk}}^{y_{jk}} \left(1 - \pi_{z|w_{jk}}\right)^{1-y_{jk}}$$

TABLE 5.11

Mean and Variance of the Number of Live Fetuses in Each Dose Group

Dose	0	0.044	0.091	0.191	0.292
Mean	13.2	12.30769	12.26923	11.12	12.29167
Variance	5.682759	6.941538	6.124615	22.44333	3.259058

where $\pi_{z|w_{jk}} = \Phi\left(\delta_{y|w_{ij}}\right)$ represents the conditional expectation of the dichotomous malformation variable, i.e. $E(Y_{ij} \mid W_{ij})$. Using specific parametric dose-response models for fetal weight, the probability of malformation, and the probability of death/resorption, the joint distribution is expressed in terms of parameters to be estimated. To avoid numerical integration in order to find the parameter estimates, one can utilize a Bayesian framework with hyper-priors and applying the Gibbs Sampler. See Tables 5.11 and 5.12.

The Bayesian methodology under a very general framework was also described by Dunson (2000). Modeling methods described in the last section for analyzing discrete and continuous outcomes from developmental toxicity studies are likelihood based and thus rely on asymptotic results of the maximum likelihood theory. Thus inferences are often based on the asymptotic normality of the estimate. Dunson (2000) argues that the Bayesian approach has several important advantages. First, by applying the MCMC methods, the exact posterior distributions of the parameters and latent variables can be estimated. In addition, the Bayesian approach has the advantage that it allows for the direct incorporation of prior knowledge and in turn avoids constraints on parameters that are often necessary to ensure identifiability. Classical methods often require that a subset of the parameters is known to ensure identifiability. Also, assigning informative priors to parameters can lead to more precise estimates. By defining a new class of latent variable models for the outcomes, Dunson (2000) derives a general framework for multiple clustered outcomes. The MCMC method is to derive the posterior distribution of the parameters of the latent variable models. A brief description of the Dunson (2000) approach is provided here.

Assume that there are l measurements on each fetus and the vector of measurements on the kth pup in the jth litter is denoted by $\mathbf{y}_{jk} = \left(y_{jk1}, \ldots, y_{jkl}\right)^T$. Assume that there is a vector of underlying variables $\mathbf{z}_{jk} = \left(z_{jk1}, \ldots, z_{jkl}\right)^T$ so that z_{jkr} corresponds to $y_{jkr}; r = 1, \ldots, l$ and that $\mathbf{y}_{jk}$ and $\mathbf{z}_{jk}$ are linked through a known function

$$\mathbf{y}_{jk} = g(\mathbf{z}_{jk}, \beta),$$

where β is the vector of parameters. Assume further that the distribution of the underlying variable z_{jkr} belongs to the exponential family,

$$f(z_{jkr}) = \exp\left\{\tau_r\left[z_{jkr}\theta_{jkr} - h(z_{jkr})\right] + a_k\left(z_{jkr}, \tau_r\right)\right\} \qquad (5.26)$$

TABLE 5.12
Developmental Toxicity of Ethylene Glycol in Mice

Dose (mg/kg)	Dams	Live		Litter Size		Malformed		Fetal Weight (g)	
		Number	Proportion	Mean	SD	Number	Proportion	Mean	SD
0	25	297	0.89	11.88	2.45	1	0.003	0.972	0.098
750	24	276	0.89	11.5	2.38	26	0.094	0.877	0.104
1500	22	229	0.86	10.41	3.46	89	38.9	0.764	0.107
3000	23	226	0.8	9.83	2.69	129	57.1	0.704	0.124

Data from Price et al. (1985) and Catalano and Ryan (1992).

where τ_r is a scalar and θ_{jkr} is the canonical parameter. Also, the mean and variance of z_{jkr} are given by

$$E(z_{jkr}) = \frac{d\,h(z_{jkr})}{d\vartheta_{jkr}}$$

and

$$\mathrm{Var}(z_{jkr}) = \frac{1}{\tau_r}\frac{d^2 h(z_{jkr})}{d\theta_{jkr}^2}.$$

Now, we use a generalized linear model with three sets of covariates to relate the canonical parameter θ_{jkr} to the latent variables

$$Q_r(\theta_{jkr}) = x_{jkr}^T \beta + u_{jkr}^T\,b_j + v_{jkr}^T c_{jk} \tag{5.27}$$

where $Q_{r(.)}$ is a monotonic link function relating the measurements on the kth pup to the covariates, x_{jkr} is the vector of covariates, u_{jkr} is the vector of covariates linking the latent variables induced by the jkth fetus to the Q_r and v_{jkr} is the vector of covariates linking the latent variables induced by measurements on the jkth fetus to Q_r. According to Dunson (2000), in equation (5.26), the first term describes the global effect of the covariates on each litter, the second term describes how the effects are modified for each pup and the third term describes how the effects are modified for the measurements on each pup. The latent variables b_j and c_{jk} are assigned either multivariate normal distributions or linked to simple exponential models; see Dunson (2000) for more details and how by assigning appropriate prior distributions on the model parameters and applying the MCMC methodology, the posterior distributions are derived. This modeling framework is very general and accommodates any type of measurements with clustered outcomes. The methodology is extended in Dunson (2003) to include the litter size as a variable as well.

We will not explore the Bayesian models any further here, except that we mention that a Bayesian nonparametric mixture modeling framework was developed by Fronczyk and Kottas (2014) and Kottas and Fronczyk (2013). The mixture model is "built from dependent Dirichlet process prior, with dependence of the mixing distribution governed by the dose level" to provide additional flexibility in the functional form.

Exercises

1. Substitute the one-hit model given in (5.2) in the beta-binomial log-likelihood given in (5.1). Differentiate with respect to the

unknown parameters and derive the likelihood equations. Describe the equations in a matrix format and comment on how to derive the maximum likelihood estimates.

2. Following Gaylor and Razzaghi (1992), assume that for a viable fetus to be normal, the embryo would have to go through n developmental stages with normal outcomes. Let Q_i; $i = 1, ..., n$ be the probability that the outcome of the ith stage is normal given that the outcome of the $(i - 1)$th stage is normal. Suppose that at a given dose D, for each stage to have a normal outcome, cell replication is required to reach a critical stage, and suppose that cell growth is modeled at each stage following a logistic model

$$\frac{d}{dt}N_i(t,D) = \theta_1(D)N_i(t,D) - \theta_2(D)\left[N_i(t,D)\right]^2.$$

Assume further that the relative probabilities of successfully producing enough cells during the ith stage to form a normal stage is equal to the ratio of the number of cells produced at dose (D) compared to the control animals ($D = 0$), i.e.

$$\frac{Q_i(D)}{Q_i(0)} = \frac{N_i(t,D) - N_0}{N_i(t,0) - N_0}.$$

a. Assuming the Weibull model

$$\theta_j(D) = \theta_{j0} \exp\left[-(\theta_{j1} D^\gamma)\right] \quad \theta_{j0}, \theta_{j1} > 0, j = 1, 2$$

for $\theta_1(D)$ and $\theta_2(D)$, derive an expression for the probability of a normal fetus at dose D.

b. For a simpler case of constants θ_1 and θ_2, fit the cleft palate data for the CD-1 strain mice from Table 5.1. Comment on the results.

3. Using the logistic model as the dose response, derive an expression for the BMD when additional risk is used as the measure of risk. Fit the logistic model to the CD-1 mice in the cleft palate data and obtain an estimate of the BMD.

4. Verify equation (5.16), that is, show that

$$\int_0^1 \int_0^1 \binom{n_{ij}}{y_{ij1}, y_{ij2}} p_{ij1}^{y_{ij1}} p_{ij2}^{y_{ij2}} p_{ij3}^{y_{ij3}} \cdot \frac{\Gamma(\alpha_i + \beta_i + \gamma_i)}{\Gamma(\alpha_i)\Gamma(\beta_i)\Gamma(\gamma_i)} p_{ij1}^{\alpha_i-1} p_{ij2}^{\beta_i-1} p_{ij3}^{\gamma_i-1} dp_{ij1} dp_{ij2}$$

$$= \frac{n_{ij}!\Gamma(\alpha_i + \beta_i + \gamma_i)\Gamma(y_{ij1} + \alpha_i)\Gamma(y_{ij2} + \beta_i)\Gamma(y_{ij3} + \gamma_i)}{y_{ij1}!y_{ij2}!y_{ij3}!\Gamma(n_{ij} + \alpha_i + \beta_i + \gamma_i)\Gamma(\alpha_i)\Gamma(\beta_i)\Gamma(\gamma_i)}$$

5. In the Dirichlet-trinomial distribution given by (5.16)

   a. Show that the marginal distribution of Y_{ij1}, the number of dead/resorbed fetuses is a beta-binomial distribution with parameters n_{ij}, α_i, and β_i.

   b. Show that the marginal distribution of Y_{ij3}, the number of normal fetuses is a beta-binomial distribution with parameters n_{ij}, γ_i, and $\alpha_i + \beta_i$.

   c. Show that the conditional distribution of Y_{ij2}, the number of malformed fetuses given Y_{ij1}, the number of dead/resorbed fetuses is a beta-binomial distribution with parameters $(n_{ij} - y_{ij1})$, β_i, and γ_i.

   Hint: See Chen et al. (1991).

6. Show that the Dirichlet-trinomial model is a member of the class of double beta-binomial distribution. That is, show that equation (5.16) can be expressed as (5.17).

7. Using equations (5.22)–(5.24), derive the likelihood equations for the parameters $\beta1$, α, and σ^2.

   Hint: See Fitzmaurice and Laird (1995).

References

Ahn, H. and Chen, J. J. (1997). Tree-structured logistic model for over-dispersed binomial data with application to modeling developmental effects. *Biometrics*, 435–55.

Anderson, D. R. (2008). Information theory and entropy. In *Model based inference in the life sciences: A primer on evidence*, 51–82.

Bailer, A. J., Noble, R. B., and Wheeler, M. W. (2005). Model uncertainty and risk estimation for experimental studies of quantal responses. *Risk Analysis: An International Journal*, 25(2), 291–99.

Bailer, A. J. and Smith, R. J. (1994). Estimating upper confidence limits for extra risk in quantal multistage models. *Risk Analysis*, 14, 1001–10.

Boos, D. D. and Brownie, C. (1986). Testing for a treatment effect in the presence of nonresponders. *Biometrics*, 191–97.

Boos, D. D. and Brownie, C. (1991). Mixture models for continuous data in dose-response studies when some animals are unaffected by treatment. *Biometrics*, 1489–504.

Buckley, E. B., Piegorsch, W. W., and West, R. W. (2009). Confidence limits on one-stage model parameters in benchmark risk assessment. *Environmental and Ecological Statistics*, 16, 53–62.

Burnham, K. P., Anderson, D. R., and Huyvaert, K. P. (2011). AIC model selection and multimodel inference in behavioral ecology: Some background, observations, and comparisons. *Behavioral Ecology and Sociobiology*, 65(1), 23–35.

Burnham, K. and Anderson, D. R. (2002). *Model selection and multimodel inference*. Springer, New York.

Calabrese, E. J. (2014). Hormesis: A fundamental concept in biology. *Microbial Cell*, 1(5), 145.

Catalano, P. J. and Ryan, L. M. (1992). Bivariate latent variable models for clustered discrete and continuous outcomes. *Journal of the American Statistical Association*, 87(419), 651–58.

Catalano, P. J., Scharfstein, D. O., Ryan, L. M., Kimmel, C. A., and Kimmel, G. L. (1993). Statistical model for fetal death, fetal weight, and malformation in developmental toxicity studies. *Teratology*, 47, 281–90.

Chen, J. (1993). A malformation incidence dose-response model incorporating fetal weight and/or litter size as covariates. *Risk Analysis*, 13(5), 559–64.

Chen, J. J. and Gaylor, D. W. (1992). Dose-response modeling of quantitative response data for risk assessment. *Communications in Statistics-Theory and Methods*, 21, 2367–81.

Chen, J. J. and Kodell, R. L. (1989). Quantitative risk assessment for teratological effects. *Journal of the American Statistical Association*, 84(408), 966–71.

Chen, J. J., Kodell, R. L., Howe, R. B., and Gaylor, D. W. (1991). Analysis of trinomial responses from reproductive and developmental toxicity experiments. *Biometrics*, 1049–58.

Chen, J. J. and Li, L.-A. (1994). Dose-response modeling of trinomial responses from developmental experiments. *Statistica Sinica*, 4, 265–74.

Conover, W. J. and Salsburg, D. S. (1988). Locally most powerful tests for detecting treatment effects when only a subset of patients can be expected to "respond" to treatment. *Biometrics*, 189–96.

Crump, K. S. (1984). A new method for determining allowable daily intakes. *Toxicological Sciences*, 4(5), 854–71.

Declerk, L., Molenberghs, G., Aerts, M., and Ryan, L. (2000). Litter-based methods in developmental toxicity risk assessment. *Environmental and Ecological Statistics*, 7(1), 57–76.

Dempster, A. P., Patel, C. M., Selwyn, M. R., and Roth, A. J. (1984). Statistical and computational aspects of mixed model analysis. *Journal of the Royal Statistical Society: Series C (Applied Statistics)*, 33(2), 203–14.

Dunson, D. B. (2000). Bayesian latent variable models for clustered mixed outcomes. *Journal of the Royal Statistical Society: Series B (Statistical Methodology)*, 62, 355–66.

Dunson, D. B. (2003). Dynamic latent trait models for multidimensional longitudinal data. *Journal of the American Statistical Association*, 98, 555–63.

Faes, C., Geys, H., Aerts, M., and Molenberghs, G. (2006). A hierarchical modeling approach for risk assessment in developmental toxicity studies. *Computational Statistics & Data Analysis*, 51(3), 1848–61.

Faes, C., Geys, H., Aerts, M., Molenberghs, G., and Catalano, P. J. (2004). Modeling combined continuous and ordinal outcomes in a clustered setting. *Journal of Agricultural, Biological, and Environmental Statistics*, 9(4), 515.

Faustman, E. M., Wellington, D. G., Smith, W. P., and Kimmel, C. A. (1989). Characterization of a developmental toxicity dose-response model. *Environmental Health Perspectives*, 79, 229–41.

Fitzmaurice, G. M. and Laird, N. M. (1995). Regression models for a bivariate discrete and continuous outcome with clustering. *Journal of the American Statistical Association*, 90, 845–52.

Fitzmaurice, G. M., Laird, N. M., and Ware, J. H. (2011). *Applied longitudinal analysis*, 2nd edition. John Wiley & Sons, Hoboken.

Fronczyk, K. and Kottas, A. (2014). A Bayesian approach to the analysis of quantal bioassay studies using nonparametric mixture models. *Biometrics*, 70, 95–102.

Gart, J. J., Krewski, D., Lee, P. N., Tarone, R. E., and Wahrendorf, J. (1986). Statistical methods in cancer research. Volume III—The design and analysis of long-term animal experiments. *IARC Scientific Publications*, 79, 1–219.

Gaylor, D. W. (1994). Dose-response modeling. In *Developmental toxicology*, 2nd edition, C. A. Kimmel and J. Buelke-Sam eds. Raven Press, New York, 363–75.

Gaylor, D. W. and Razzaghi, M. (1992). Process of building biologically based dose–response models for developmental defects. *Teratology*, 46, 573–81.

Gaylor, D. W., Sheehan, D. M., Young, J. F., and Mattison, D. R. (1988). The threshold dose question in teratogenesis. *Teratology*, 38(4), 389–91.

Hale, F. (1937). The relation of maternal vitamin A deficiency to microphthalmia in pigs. *Texas State Journal of Medicine*, 33, 228–23.

Hall, D. B. and Severini, T. A. (1998). Extended generalized estimating equations for clustered data. *Journal of the American Statistical Association*, 93, 1365–75.

Hoetling, J. A., Madigan, D., Raftery, A. E., and Volinsky, C. T. (1999). Bayesian model averaging: A tutorial. *Statistical Science*, 14(4), 382–401.

Holson, J. F., Gaines, T. B., Nelson, C. J., LaBorde, J. B., Gaylor, D. W., Sheehan, D. M., and Young, J. F. (1992). Developmental toxicity of 2, 4, 5-trichlorophenoxyacetic acid (2, 4, 5-T) I. Multireplicated dose-response studies in four inbred strains and one outbred stock of mice. *Toxicological Sciences*, 19(2), 286–97.

Hunt, D. L., Rai, S. N., and Li, C. S. (2008). Summary of dose-response modeling for developmental toxicity studies. *Dose-Response*, 6(4), dose-response.

Kang, S. H., Kodell, R. L., and Chen, J. J. (2000). Incorporating model uncertainties along with data uncertainties in microbial risk assessment. *Regulatory Toxicology and Pharmacology*, 32(1), 68–72.

Khorsheed, E. and Razzaghi, M. (2019). Bayesian model averaging for benchmark dose analysis in developmental toxicology. *Applied Math*, 13(1), 1–10.

Kimmel, C. A. and Gaylor, D. W. (1988). Issues in qualitative and quantitative risk analysis for developmental toxicology 1. *Risk Analysis*, 8, 15–20.

Kottas, A. and Fronczyk, K. (2013). Flexible Bayesian modelling for clustered categorical responses in developmental toxicology. In *Bayesian Theory and Application*, P. Damien, P. Dellaportas, N. G. Polson, and D. A. Stephens (eds.). Oxford University Press, Oxford, 70–83.

Kupper, L. L., Portier, C., Hogan, M. D., and Yamamoto, F. (1986). The impact of litter effects on dose response modeling in teratology. *Biometrics*, 42, 85–98.

Leisenring, W. and Ryan, L. (1992). Statistical properties of the NOAEL. *Regulatory Toxicology and Pharmacology*, 15, 161–71.

Liang, K. Y. and Hanfelt, J. (1994). On the use of the quasi-likelihood method in teratological experiments. *Biometrics*, 50(3), 872–80.

Liang, K. Y. and Zeger, S. L. (1986). Longitudinal data analysis using generalized linear models. *Biometrika* 73, 13–22.

Liang, K. Y., Zeger, S. L., and Qaqish, B. (1992). Multivariate regression analyses for categorical data. *Journal of the Royal Statistical Society: Series B (Methodological)*, 54(1), 3–24.

Mack, G. A. and Wolfe, D. A. (1981). K-sample rank tests for umbrella alternatives. *Journal of the American Statistical Association*, 76(373), 175–81.

Mattison, D. R., Wilson, S., Coussens, C., and Gilbert, D. (Eds.) (2003). *The role of environmental hazards in premature birth.* The National Academies Press, Washington, DC.

Mohlenberghs, G. and Ryan, L. M. (1999). An exponential family model for clustered multivariate binary data. *Environmetrics: The Official Journal of the International Environmetrics Society,* 10, 279–300.

Moon, H., Kim, H. J., Chen, J. J., and Kodell, R. L. (2005). Model averaging using the Kullback information criterion in estimating effective doses for microbial infection and illness. *Risk Analysis: An International Journal,* 25(5), 1147–59.

Piegorsch, W. W., An, L., Wickens, A. A., Webster West, R., Peña, E. A., and Wu, W. (2013). Information-theoretic model-averaged benchmark dose analysis in environmental risk assessment. *Environmetrics,* 24, 143–57.

Prentice, R. L. (1986). Binary regression using an extended beta-binomial distribution, with discussion of correlation induced by covariate measurement errors. *Journal of the American Statistical Association,* 81(394), 321–27.

Price, C. J., Kimmel, C. A., George, J. D., and Marr, M. C. (1987). The developmental toxicity of diethylene glycol dimethyl ether in mice. *Fundamental and Applied Toxicology,* 8, 115–26.

Price, C. J., Kimmel, C. A., Tyl, R. W., and Marr, M. C. (1985). The developmental toxicity of ethylene glycol in rats and mice. *Toxicology and Applied Pharmacology,* 81(1), 113–27.

Raftery, A. E. (1995). Bayesian model selection in social research. *Sociological Methodology,* 25, 111–63.

Rai, K. and Van Ryzin, J. (1985). A dose-response model for teratological experiments involving quantal responses. *Biometrics,* 41, 1–9.

Razzaghi, M. (1999). Quantitative risk assessment for developmental toxicants in nonhomogeneous populations. In *Statistics for the environment, Volume 4, Pollution assessment and control,* V. Barnett, A. Stein, and K. F. Turkman eds. John Wiley, New York, 209–21.

Razzaghi, M. and Loomis, P. (2001). The concept of hormesis in developmental toxicology. *Human and Ecological Risk Assessment,* 7, 933–42.

Razzaghi, M. and Nanthakumar, A. (1992). On using Lehmann alternatives with nonresponders. *Mathematical Biosciences,* 109, 69–83.

Razzaghi, M. and Nanthakumar, A. (1994). A locally most powerful test for detecting a treatment effect in the presence of nonresponders. *Biometrical Journal,* 36, 373–84.

Regan, M. M. and Catalano, P. J. (1999). Bivariate dose-response modeling and risk estimation in developmental toxicology. *Journal of Agricultural, Biological, and Environmental Statistics,* 4(3), 217–37.

Ryan, L. (1992a). Quantitative risk assessment for developmental toxicity. *Biometrics,* 48(1), 163–74.

Ryan, L. (1992b). The use of generalized estimating equations for risk assessment in developmental toxicity. *Risk Analysis,* 12(3), 439–47.

Ryan, L. M., Catalano, P. J., Kimmel, C. A., and Kimmel, G. L. (1991). Relationship between fetal weight and malformation in developmental toxicity studies. *Teratology,* 44(2), 215–23.

Salsburg, D. (1986). Alternative hypotheses for the effects of drugs in small-scale clinical studies. *Biometrics,* 43(3), 671–74.

Shao, K. and Small, M. J. (2011). Potential uncertainty reduction in model-averaged benchmark dose estimates informed by an additional dose study. *Risk Analysis: An International Journal*, 31(10), 1561–75.

Simmons, S. J., Chen, C., Li, X., Wang, Y., Piegorsch, W. W., Fang, Q., Hu, B., and Dunn, G. E. (2015). Bayesian model averaging for benchmark dose estimation. *Environmental and Ecological Statistics*, 22(1), 5–16.

Tyl, R. W. and Marr, M. C. (2006). Developmental toxicity testing—Methodology. In *Developmental and Reproductive Toxicology*. CRC Press, 153–97.

Tyl, R. W., Price, C. J., Marr, M. C., Myers, C. B., Seely, J. C., Heindel, J. J., and Schwetz, B. A. (1993). Developmental toxicity evaluation of ethylene glycol by gavage in New Zealand white rabbits. *Toxicological Sciences*, 20(4), 402–12.

US EPA. (1991). Guidelines for developmental toxicity risk assessment. *Federal Register*, 56(234), 63798–826, EPA/600/FR-91/001.

Wasserman, L. (2000). Bayesian model selection and model averaging. *Journal of Mathematical Psychology*, 44, 92–107.

Wedderburn, R. W. (1974). Quasi-likelihood functions, generalized linear models, and the Gauss—Newton method. *Biometrika*, 61(3), 439–47.

West, R. W., Piegorsch, W. W., Peña, E. A., An, L., Wu, W., Wickens, A. A., Xiong, H., and Chen, W. (2012). The impact of model uncertainty on benchmark dose estimation. *Environmetrics*, 23(8), 706–16.

Wheeler, M. W. and Bailer, A. J. (2007). Properties of model-averaged BMDLs: A study of model averaging in dichotomous response risk estimation. *Risk Analysis: An International Journal*, 27(3), 659–70.

Wheeler, M. W. and Bailer, A. J. (2013). An empirical comparison of low-dose extrapolation from points of departure (PoD) compared to extrapolations based upon methods that account for model uncertainty. *Regulatory Toxicology and Pharmacology*, 67(1), 75–82.

Wheeler, M. and Bailer, A. J. (2012). Monotonic Bayesian semiparametric benchmark dose analysis. *Risk Analysis: An International Journal*, 32(7), 1207–18.

Whitney, M. and Ryan, L. (2009). Quantifying dose-response uncertainty using bayesian model averaging. In *Uncertainty modeling in dose response: bench testing environmental toxicity*, R. M. Cooke (ed.). John Wiley & Sons, Hoboken, 165–79.

Williams, D. A. (1975). 394: The analysis of binary responses from toxicological experiments involving reproduction and teratogenicity. *Biometrics*, 31(4), 949–52.

Williams, D. A. (1982). Extra-binomial variation in logistic linear models, *Applied Statistics*, 31, 144–48.

Williams, P. L. and Ryan, L. M. (1996). Dose-response models for developmental toxicology. In *Handbook of developmental toxicology*, R. D. Hood ed. CRC Press, New York, 635–66.

6

Analysis of Quantitative Data and Responses from Neurotoxicity Experiments

6.1 Introduction

Historically, most health risk estimation techniques for exposure to toxic chemicals have focused on quantal data primarily because many of the studies concentrated on the carcinogenic effects of substances. Thus a large number of animal bioassay studies focused on estimating the risk of the occurrence of cancer by classifying animals as being with or without a tumor. Even, in developmental toxicity studies, as we saw in Chapter 5, a large body of literature was published for modeling the probability of incidence of a malformation. However, in many experiments, the measurement of interest is non-quantal and continuous in nature. For example, as we saw in Chapter 5, the weight of a fetus at birth can directly be affected by maternal exposure to a toxic substance during pregnancy.

Other examples of continuous outcomes include hematological, biochemical, neurological, and behavioral measurements. For this reason, interest in the development of models that concentrate on continuous outcomes has grown exponentially since the early 1990s. In this chapter, we discuss various dose-response models that have been developed for continuous outcomes. Specifically, later in this chapter, we consider how some of the models have been applied for estimating the risk in experiments that study the adverse neurological effects of chemicals.

6.2 Defining an Abnormal Response

Note that unlike quantal data where there is a distinct demarcation for adverse effect, in continuous outcomes there is no clear definition of how an adverse effect is defined. For example in carcinogenicity experiments an animal is classified as either having cancer or free of cancer. Similarly, in developmental toxicity experiments, an embryo can be classified as either having a malformation or being free of such deformity. For quantitative responses, however, such distinction between normal and abnormal responses is not available. In the absence of a clear criterion for defining an adverse effect in continuous outcomes, several authors have suggested that an adverse effect in exposed animals defines an adverse effect as a response falling outside the normal range observed in control unexposed animals. In other words, an outcome is considered to be abnormal if it lies in the direction of adversity and far enough from the control mean that its occurrence is highly unusual and unlikely in control animals. One approach in specifying such a response is to let an abnormal response to be a response that falls beyond a certain number of standard deviations from the mean response of the unexposed animals. Denoting the quantitative response of interest at dosage d by $X(d)$, then if, for example, the direction of adversity of response is to the left of the control mean, the abnormal response may be defined as a response for which $X(d) < \mu(0) - k\sigma(0)$ where $\mu(d)$ and $\sigma(d)$ respectively denote the mean and standard deviation of responses at the exposure level d. Thus,

$$P(\text{adverse outcome}) = P[X(d) < \mu(0) - k\sigma(0)] \tag{6.1}$$

Often the value of k is chosen to be 2.33 which corresponds to the 1% tail area of the normal distribution or more commonly 3. Clearly the choice of k can play a very important role in defining an adverse effect and abnormal response. Later in this chapter, we will discuss the impact of the choice of k on risk assessment. An alternative approach to define a cut-off value for an adverse effect in quantitative responses is to fix a small tail probability in unexposed animals and find the corresponding response. Thus, a response $X(d)$ is abnormal if $X(d) < c_0$ where

$$P[X(0) < c_0] = \alpha_0$$

where α_0 is a fixed predetermined value such as 0.01 or 0.001. Note that in the above definition of adverse effect, it is assumed that abnormal response in treated animals at dose d is a response that is on the lower tail of the response distribution of the control animals. Equivalently, if high responses are considered to be abnormal, we could use the upper tail of the distribution of the responses of the unexposed animals to indicate an abnormal response. Figure 6.1 illustrates the definition of adverse effect when the distribution of both the exposed and unexposed individuals are normal.

6.3 Risk Assessment for Continuous Responses

With the above definition of an abnormal response, the risk in exposed animals is therefore the area under the curve beyond the cut-off value for

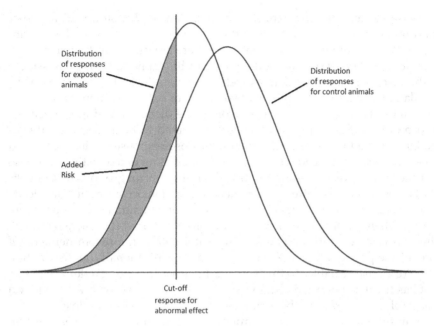

Distribution
of responses
for exposed
animals

Distribution
of responses
for control animals

Added
Risk

Cut-off
response for
abnormal effect

FIGURE 6.1
Definition of Adverse Effect for Continuous Responses.

abnormality and the excess risk is simply the difference in the areas under both the exposed and unexposed animals beyond the point of abnormality. Also, as before, to determine a safe exposure level, the so-called hybrid approach is used where a small nominal value of additional risk called benchmark risk (BMR) is fixed and the dose corresponding to that risk called benchmark dose (BMD) is calculated. However, as discussed in previous chapters, the point estimate of risk at a given dose or the benchmark dose is highly model dependent and often not reliable. Similar to what we saw for quantal response data, an upper confidence limit for risk is generally calculated. Additionally, a lower confidence for the benchmark dose (BMDL) is determined as a more reliable estimate of a safe exposure level.

Calculation of risk or the benchmark dose for continuous data therefore requires knowledge of the distribution of the underlying continuous response variable, which may not be available. The use of the normal or the lognormal distributions is very common in this respect. But, as we will see later in this chapter, if the distribution is misspecified, the normal or the lognormal distribution is used when the data or its logarithm exhibits skewness; we may get unreliable estimates for the risk or the benchmark dose. In fact, even if the responses for the animals in the control unexposed group can be assumed to be normal, the distribution can change in the exposed individuals as exposure might affect some animals more than others. One approach to resolve this problem is through quantalization of the continuous response variable. Once a cut-off value for an adverse effect is defined, each response variable for the exposed animals can be dichotomized as either abnormal if it falls beyond the cut-off value or normal otherwise. The advantage of this dichotomization is that since there are direct methods for quantal response data, those methodologies can be applied for risk assessment in the case of continuous data as well. However, through a comprehensive simulation, Gaylor (1996) studies the problem of risk assessment for continuous data in two cases when the outcomes are dichotomized and when the distribution of the response variable is assumed to be a simple nonlinear model. Using a multi-dose experiment, the dose-response model is assumed to be log-linear and an adverse effect is defined as an outcome that occurs so that the proportion of abnormal animals in the control group is 0.01 or 0.05. The conclusion is that although the dichotomization of continuous data leads to more direct risk assessment, the estimates of benchmark doses are more precise when the actual quantitative responses are used for risk assessment. Thus converting continuous data to quantal data is not recommended. Of course, one problem that still remains is the specification of the distribution of the response variable. To this end, it is believed that using a flexible model that lets the data choose the shape of the model at each dose level can play a major role in accuracy of the estimates of risk and benchmark doses.

To find the upper confidence limit on risk, the same two approaches that were discussed for quantal data also apply here. Accordingly, one approach is to use the asymptotic normality of maximum likelihood estimates of the model parameters. The second approach is to apply the asymptotic distribution of the likelihood ratio statistic. In the latter approach, to find BMD and BMDL, the usual method is to invert the risk equation. Both approaches have interchangeably been used in publications relating to this issue, although Banga et al. (2002) find that the approach based on asymptotic distribution of the likelihood ratio statistic provides more reliable coverage when responses are not normally distributed. If the responses are normally distributed, Banga et al. (2001) find through simulation that the two methods are effectively indistinguishable.

6.4 Common Dose-Response Models

In a draft report by the World Health Organization (WHO) (2008) some of the dose-response models for continuous responses are listed. Table 6.1 provides a summary of these models.

Each of these models can be combined with a statistical distribution such as normal or lognormal to describe the relationship between the exposure level and a quantitative response in a population, with the given model corresponding to the central estimate. One of the earliest attempts to characterize model-based risk for continuous outcomes is by Gaylor and

TABLE 6.1

Common Dose-Response Models for Continuous Data

Name	Equation	Notes
Michaelis-Menton	$\dfrac{R_{Max}\{S\}}{K_M + [S]}$	R_{Max} = Maximum Reaction rate $[S]$ = Substrate Concentration K_M = Michaelis-Menton Constant
Hill Equation	$\dfrac{R_{Max}D^n}{K_D^n + D^n}$	R_{Max} = Maximum Response D = Dose K_D = Reaction Constant n = Number of Binding Sites
Exponential	$R_{Max}[1 - e^{-tD}]$	R_{Max} = Maximum Response D = Dose t = Exponential Rate
Power	βD^t	β = Scale Parameter D = Dose t = Shape Parameter
Linear	$\beta_0 + \beta_1 D$	β_0 = Intercept β_1 = Slope
Polynomial	$\displaystyle\sum_{r=0}^{l} \beta_r D^r$	$\beta_0, \ldots, \beta_l$ = Constants

Source: Adapted from WHO (2008).

TABLE 6.2

Developmental Effect of Diethylhexyl Phthalate (DEHP) in Mice

Dose	Number Treated	Mean	SD
0	6	22.57	5.61
0.1	6	18.54	9.87
0.3	6	3.88	4.09
0.6	6	2.57	2.77
1.2	6	3.7	3.11

Source: Data from Tyl et al. (1983).

Slikker (1990). The logarithm of the mean response is modeled as a quadratic function of dose. Defining an abnormal adverse effect to be a response that occurs rarely with a small probability in the control animals as in (6.1), and using a log-quadratic dose-response model with the lognormal distribution for the responses, the point estimate of risk is derived at a given dose.

Example 6.1

To illustrate, consider the data in Table 6.2 which give the average maternal weight gain along with the associated standard deviation at each dose level during the gestation period for diethylhexyl phthalate (DEHP) in CD-1 mice (data from Tyl et al., 1983). Let $y(d)$ be the average weight at exposure level d and assume

TABLE 6.3

Parameter Estimates for Fitting Equation 6.2

Parameter	Estimate	Standard Deviation	t-value	P-value
Intercept	3.231	0.258	12.523	0.00632
Linear	−6.479	1.259	−5.147	0.0357
Quadratic	4.076	0.996	4.093	0.0548

$$\log[y(d)] = \beta_0 + \beta_1 d + \beta_2 d^2 \tag{6.2}$$

Using the data from Table 6.2 to fit the above model, we find that the parameter estimates are:

$$\widehat{\beta_0} = 3.2310, \ \widehat{\beta_1} = -6.4792, \text{ and } \widehat{\beta_2} = 4.0758.$$

Table 6.3 provides a summary of the results of fitting (6.2). The fitted model is significant ($p = 0.051$) with $R^2 = 0.95$. Suppose now that it is desired to estimate the risk at dose $d = 0.025$. From (6.2), the estimate of the average weight gain at this dose level is 21.57 g. Define an abnormal response to be an outcome that falls three or more standard deviation below the mean of control animals, that is, assume that $k = 3$ in (6.1). Note that from Table 6.3 and from (6.2) the estimates of the mean and standard deviation of the responses for control animals are $e^{\widehat{\beta_0}} = 25.305$ and $e^{\text{sd}(\widehat{\beta_0})} = 1.294$. Now, in Table 6.2, the estimated cut-off average weight for an abnormal response is given by

$$\log[\dot{y}(0)] - 3\hat{\sigma}_0 = \widehat{\beta_0} - 3\text{sd}(\widehat{\beta_0}) = 3.231 - 3(.258) = 2.457.$$

Thus maternal weight gain is considered to be abnormal if it falls below $\exp(2.457) = 11.670$ g. If we assume a lognormal distribution for the weight of control animals, then the probability of a control animal having a weight of 11.670 g or less is $P(Z < -3) = 0.00135$, which clearly is a rare event in control animals. Now, at the desired dose of $d = 0.025$, the estimate of risk is given by

$$\hat{\pi}(0.025) = P\left(Z < \frac{\log(11.670) - \log(21.57)}{\log(1.294)} \right) = 0.00858$$

where as usual Z is the standard normal variable. We can also compute an approximate upper confidence level on risk using the asymptotic normality of the logarithm of responses. For example if $u_{0.95}$, an approximate upper 95% confidence interval is desired, we have

$$\log(u_{0.95}) = \log \hat{\pi}(0.025) + 1.645\sqrt{\text{Var}(\log y)}$$

TABLE 6.4

Covariance Matrix of Parameter estimates

	Intercept	Linear	Quadratic
Intercept	0.0666	−0.2379	0.155
Linear	−0.2379	1.5844	−1.2084
Quadratic	0.155	−1.2084	0.9914

We can find an estimated Var (log y) by using (6.2). We have

$$\text{Var}(\log[y(d)]) = \text{Var}\left(\widehat{\beta_0}\right) + d^2 \, \text{Var}\left(\widehat{\beta_1}\right) + d^4 \, \text{Var}\left(\widehat{\beta_2}\right)$$
$$+ 2d\,\text{Cov}\left(\widehat{\beta_0},\widehat{\beta_1}\right) + 2d^2 \, \text{Cov}\left(\widehat{\beta_0},\widehat{\beta_2}\right)$$
$$+ 2d^3 \, \text{Cov}\left(\widehat{\beta_1},\widehat{\beta_2}\right).$$

Table 6.4 gives the variance–covariance matrix of the parameter estimates from fitted model (6.2). Substituting the results in the above equation, we find,

$$\widehat{\text{Var}}(\log[y(0.025)]) = 0.05586$$

Thus, a 95% upper confidence limit on risk at dose $d=0.025$ is given by

$$\hat{u}_{0.95} = \exp\left\{\log 0.00858 + 1.645\sqrt{0.05586}\right\} = 0.00217$$

Similarly, it is also possible to estimate a dose level that will result in no more than a fixed low acceptable level of risk. Assume for example, that a risk of 1%, i.e. $\pi_0 = 0.01$, is considered acceptable in this case. Then using the percentiles of a standard normal distribution, from (6.2), we need to solve

$$2.33 = \frac{3.2310 - 6.4749\,d + 4.0758\,d^2 - \log(11.670)}{\log(1.294)}$$

which results in a benchmark dosage level of BMD $= 0.054$ g/day. We can also determine a BMDL by applying the asymptotic theory as well, i.e. using

$$\text{BMDL} = \text{BMD} - 1.645\sqrt{\text{Var(BMD)}}$$

for which we first need to obtain an estimate for the Var (BMD). One way to accomplish this task is by first solving the equation

$$2.33 = \frac{\beta_0 + \beta_1 d + \beta_2 d^2 - \log 11.670}{\log 1.294}$$

to find an equation for BMD as

$$\text{BMD} = \left(2\hat{\beta}_2\right)^{-1}\left\{-\hat{\beta}_1 + \left[\hat{\beta}_1 - 4\hat{\beta}_2\left(\hat{\beta}_0 - 3.0576\right)\right]^{1/2}\right\} \qquad (6.3)$$

and using Taylor expansion to find an approximate value for Var (d). This approach may be too cumbersome and not very accurate. Alternatively, let $\beta = (\beta_0, \beta_1, \beta_2)^T$. Then using the delta method (see e.g. Casella and Berger, 2002) we have

$$\text{Var(BMD)} = \left[\frac{\partial\,\text{BMD}}{\partial\beta}\right]^T \text{Var}\hat{\beta}\left[\frac{\partial\,\text{BMD}}{\partial\beta}\right] \qquad (6.4)$$

where BMD is given by (6.3); see Exercise 1 at the end of this chapter.

6.4.1 Normal Distribution

Although the above procedure leads to useful estimates for the risk and the benchmark dose, it cannot be very reliable since firstly the modeling is on the average responses and secondly it is not easily amenable to finding upper confidence levels for risk and lower confidence bounds for the benchmark dose. Early attempts in modeling the individual responses was proposed by Kodell and West (1993) and West and Kodell (1993) where a normal distribution,

$$P(d) = \frac{1}{\sigma\sqrt{2\pi}}\,e^{-\frac{1}{2\sigma^2}\{X(d)-\mu(d)\}^2} \qquad (6.5)$$

with $X(d)$ representing the individual continuous response at dosage d, was used as the dose-response relationship. Using the same definition for an adverse effect, from (6.1) and (6.5) we have,

$$P(\text{adverse outcome}) = P\left\{\frac{X(d)-\mu(d)}{\sigma_d} < \frac{\mu(0)-\mu(d)-k\sigma_0}{\sigma_d}\right\}$$

$$= \Phi\left(\frac{\mu(0)-\mu(d)}{\sigma_d} - k\frac{\sigma_0}{\sigma_d}\right)$$

Using the excess risk as the measure of additional risk at exposure level d, we have

$$\pi(d) = P\left[X(d) < \mu(0) - k\sigma_0\right] - P\left[X(0) < \mu(0) - k\sigma_0\right]$$

$$= \Phi\left(\frac{\mu(0)-\mu(d)}{\sigma_d} - k\frac{\sigma_0}{\sigma_d}\right) - \Phi\left(-k\frac{\sigma_0}{\sigma_d}\right) \qquad (6.6)$$

Now, if the experiment consists of a control and g non-zero dose levels

$$0 = d_0 < d_1 < \cdots < d_g,$$

then we can estimate the mean μ_i and standard deviation σ_i of responses at dose level d_i; $i = 0, 1, \ldots, g$ by maximizing the log-likelihood

$$\log L = -\sum_{i=0}^{g}\sum_{j=1}^{n_i}\left\{\left[\frac{X_j(d_i) - \mu_i}{\sigma_i}\right]^2 - \log\left[\sigma_i\sqrt{2\pi}\right]\right\} \tag{6.7}$$

where $X_j(d_i)$ denotes the response of the jth animal at the ith dose level. Clearly, in the usual way, after differentiating with respect to μ_i and σ_i and solving the likelihood equations we get

$$\widehat{\mu_i} = \frac{\sum_{j=1}^{n_i} X_j(d_i)}{n_i} \quad i = 0, 1, \ldots, g \tag{6.8}$$

and

$$\hat{\sigma}_i^2 = \frac{\sum_{j=1}^{n_i}\left(X_j(d_i) - \widehat{\mu_i}\right)^2}{n_i} \tag{6.9}$$

By using (6.5), we can determine a point estimate of risk at experimental dosage levels. The procedure does not allow interpolation or extrapolation for estimating risk at other dosage levels. Equivalently, upper confidence limits for risk can only be derived at the experimental dosage levels. Applying one of the three methods discussed in Chapter 5 e.g. the asymptotic distribution of the likelihood ratio statistic, we can substitute (6.8) and (6.9) in (6.7) to find the maximum of log L. In addition, the likelihood (6.7) is maximized subject to (6.5) to determine the maximum of log-likelihood under the risk constraint and the upper confidence limit can be determined by using the percentiles of a χ^2 distribution.

To allow extrapolation and estimation of risk and benchmark at exposure levels other than the experimental doses, Kodell and West (1993) assume common equal variances at all dose levels

$$\sigma_0^2 = \sigma_1^2 = \cdots = \sigma_g^2 = \sigma^2$$

and suggest modeling the mean $\mu(d)$ as a quadratic function of dose

$$\mu(d) = \beta_0 + \beta_1 d + \beta_2 d^2 \tag{6.10}$$

Then, the log-likelihood (6.7) becomes

$$\log L = -\sum_{i=0}^{g}\sum_{j=1}^{n_i}\left\{\left[\frac{X_j(d_i)-\beta_0-\beta_1 d-\beta_2 d^2}{\sigma}\right]^2-\log\left[\sigma\sqrt{2\pi}\right]\right\} \qquad (6.11)$$

and the risk equation (6.3) reduces to

$$\pi(d)=P\left[X(d)<\mu(0)-k\sigma\right]-P\left[X(0)<\mu(0)-k\sigma\right]$$

$$=\Phi\left(\frac{-\beta_1 d-\beta_2 d^2}{\sigma}-k\right)-\Phi(-k) \qquad (6.12)$$

Upon differentiating (6.8) with respect to the parameters β_0, β_1, β_2, and σ^2 the maximum likelihood estimates can be determined and a point estimate of risk can be derived by using (6.10) at any given dose. To derive an upper confidence bound for risk at a given dose D, we first need to find the maximum likelihood estimates subject to (6.13). To accomplish this task, we may first solve (6.13) in terms of one of the parameters, substitute in the log-likelihood equation and maximize the log-likelihood with respect to the other parameters. For example, if we solve (6.12) for β_2, we get

$$\beta_2=-\frac{\sigma}{D^2}\left\{\Phi^{-1}\left[\pi(D)+\Phi(-k)\right]+k\right\}-\frac{\beta_1}{D} \qquad (6.13)$$

which is then substituted in (6.11) in order to find the maximum likelihood estimates subject to the risk constraint. To derive the upper confidence for risk at a given dose D, Kodell and West (1993) propose a numerical procedure where an initial estimate of risk at dose D is first found by using (6.12) based on the parameter estimates from the unrestricted maximization of the log-likelihood. This value is incremented by a small quantity to find an initial value for the upper bound on risk $\pi_u(D)$. Using this initial value, the maximum of the log-likelihood is determined subject to the risk constraint. If twice the difference of the maximum values of the unrestricted and restricted log-likelihood values is not equal to the $100(1-\alpha)$th percentile of the χ^2 distribution with one degree of freedom, $\pi_u(D)$ is incremented again and the procedure is continued until equality is obtained.

Since the assumption of equality of variance at all experimental doses may be too restrictive, West and Kodell (1993) relax this assumption and estimate the risk at experimental doses and propose a method for extrapolating to low doses by using linear splines.

6.4.2 Skewed Distributions

Often, in practice, it may happen that the data lack symmetry and the normal distribution does not fit the data adequately. In such cases, using the normal dose-response model may lead to erroneous results. In fact, Banga et al. (2000) examined the sensitivity of the risk assessment to model mis-specification and found that when the data exhibit non-normality, then the calculation of upper confidence limit on risk based on a misspecified normal distribution can lead to very poor coverage. Log tranformation of the data also did not improve the coverage significantly. Thus, if the data are clearly skewed, one should consider an alternative approach. Suppose, for simplicity, the dose-response relationship is modeled as an exponential distribution, so that if $Y_j(d_i) \, j = 1,\ldots, n_i \, i = 0,\ldots, g$ denotes the jth observation in the ith dose group, assume that

$$P\left[Y_j(d_i) \le y \right] = 1 - e^{-y/\mu(d)}$$

where $\mu(d)$ is the mean function. Assume that the logarithm of the mean function $\log \mu(d)$ can be expressed as a polynomial in d and since in practice a polynomial of degree 2 often suffices, we assume that $\log \mu(d)$ is a quadratic in d as expressed in equation (6.10). Thus,

$$\log \mu(d) = \beta_0 + \beta_1 d + \beta_2 d^2.$$

Then, log-likelihood is given by

$$\log L = \sum_{i=0}^{g} \sum_{j=1}^{n_i} \log \left[e^{-\left(\beta_0 + \beta_1 d_i + \beta_2 d_i^2\right)} e^{-Y_j(d_i)/e^{\left(\beta_0 + \beta_1 d_i + \beta_2 d_i^2\right)}} \right]$$

$$= -\sum_{i=0}^{g} \sum_{j=1}^{n_i} \left[\left(\beta_0 + \beta_1 d_i + \beta_2 d_i^2\right) + Y_j(d_i) e^{-\left(\beta_0 + \beta_1 d_i + \beta_2 d_i^2\right)} \right] \tag{6.14}$$

from which the maximum likelihood estimate of the parameters can be determined. Upon setting the derivative of (6.14) with respect to β_0, β_1, and β_2 = 0, we obtain the following system of equations

$$\sum_{i=0}^{g} \sum_{j=1}^{n_i} Y_j(d_i) d_i^r \, e^{-\left(\beta_0 + \beta_1 d_i + \beta_2 d_i^2\right)} = \sum_{i=0}^{g} \sum_{j=1}^{n_i} d_i^r \quad r = 0,1,2 \tag{6.15}$$

which must be solved using an iterative procedure. Banga et al. (2002) consider a more general problem by assuming a gamma distribution for the responses. They show that the likelihood equations yield a unique solution. To define an adverse effect, since we have a skewed and non-symmetric distribution, rather than using a cut-off value that is a on the tail of the control distribution that is a certain number of standard deviations below (above) the control mean, it makes more sense to use an equivalent definition and define an abnormal response as that which occurs with a small nominal probability α in control animals. As noted by Banga et al. (2002), separate developments are needed depending on whether the direction of adversity is to the left or right of the mean of the control distribution although there is much similarity between both cases. For illustration, let us suppose the direction of adversity is to the left of the mean of controls so that a response $Y(d)$ at dosage d is considered abnormal when it falls on the lower δth percentile of the control distribution. Thus $Y(d)$ is abnormal when it falls below a threshold T where

$$\delta = P\left[Y(d) \leq T \mid d = 0\right] = 1 - e^{-T e^{-\beta_0}} \tag{6.16}$$

Solving for T, we get

$$T = -\log(1-\alpha) \cdot e^{-\beta_0}.$$

Therefore the probability of an abnormal response at dosage d is given by

$$P(d) = P\left[Y(d) \leq T\right] = 1 - e^{-T e^{-\left(\beta_0 + \beta_1 d + \beta_2 d^2\right)}} = 1 - [1-\delta]^{e^{-\left(\beta_1 d + \beta_2 d^2\right)}} \tag{6.17}$$

Now, if we use the added risk as the measure of risk over the background, then at a given dose D, the added risk is given by

$$\pi(D) = P(D) - P(0) = (1-\delta) - (1-\delta)^{e^{-\left(\beta_1 D + \beta_2 D^2\right)}} \tag{6.18}$$

Thus for the purpose of risk assessment, the point estimate of risk at dosage D is determined from (6.18) by substitution of the maximum likelihood estimates derived from (6.15). To find an upper confidence limit for risk at the given dose D, we seek the maximum value of $\pi(D)$ that satisfies (6.18) and

$$2\left[\max(\log L) - \max(\log L_1)\right] = 2\chi^2_{1,2\alpha} \tag{6.19}$$

where $\max(\log L)$ is the maximum of the unconstrained log-likelihood obtained from (6.14) by substituting the maximum likelihood of the parameters, and $\max(\log L_1)$ is the maximum of the log-likelihood subject to the risk constraint (6.18). Although this problem can be solved by using the Lagrange multiplier technique, a more direct approach is to solve (6.18) in terms of one parameter and substitute in the likelihood equations. For example, if we solve (6.18) in terms of β_2, we get

$$\beta_2(\beta_1) = -\frac{1}{D^2}\left\{ \frac{\log\left[(1-\alpha) - \pi(D)\right]}{\log(1-\alpha)} + \beta_1 D \right\} \tag{6.20}$$

By substituting in (6.14) and differentiating with respect to β_0 and β_1 we get the two equations

$$\sum_{i=0}^{g}\sum_{j=1}^{n_i} Y_j(d_i)e^{-\left(\beta_0 + \beta_1 d_i + \beta_2(\beta_1)d_i^2\right)} = N$$

$$\sum_{i=0}^{g}\sum_{j=1}^{n_i} Y_j(d_i)d_i\left(1-\frac{d_i}{D}\right)e^{-\left(\beta_0 + \beta_1 d_i + \beta_2 d_i^2\right)} = \sum_{i=0}^{g}\sum_{j=1}^{n_i} d_i\left(1-\frac{d_i}{D}\right)$$

where $N = \sum_{i=0}^{g} n_i$ is the total number of measurements. Solving the above system of equations, the restricted maximum log-likelihood in (6.19) can be determined. To find the upper confidence bound on risk, the procedure described before for normal distribution by Kodell and West (1993) may be applied. For the more general case of the gamma dose-response model with a fixed shape parameter, Banga et al. (2002) show that that the asymptotic likelihood-based upper and lower confidence limits on risk are solutions of the Lagrange equations associated with a constrained optimization problem. Through simulation, they show that the upper confidence limit on risk based on the asymptotic distribution of the likelihood ratio has coverage that is close to nominal levels unless the total sample size N is smaller than 25.

6.5 Characterization of Dose-Response Models

The general characteristics of a dose-response model for continuous outcomes discussed by Slob (2000) can be summarized as follows:

a. Continuous measurements are generally nonnegative and models allowing nonnegative responses may be unrealistic even when the fitted model falls in the negative territory outside the range of observation.

b. Models with threshold should be excluded since in the case of continuous outcomes that would imply that all individual animals have exactly the same dose threshold for the specific endpoint, which is highly unlikely. In addition, estimation of the threshold in the parametric model can create computational problems as it could be highly dependent on the starting value. The model should be sufficiently flexible to allow for a threshold-like dose-response relationship.

c. The model should be equally sensitive between individuals, i.e. have the same relative sensitivity. By relative sensitivity, it is meant that at a given dose, the ratio of responses of the exposed and unexposed individuals is the same across species and sexes. To account for sensitivity, a potency factor is suggested. Thus, the family of models should be of the form

$$y(d) = \alpha f\left(\frac{d}{\beta}\right)$$

where y is the response, d is the administered dose, and a and b are respectively parameters representing the sensitivity and potency.

d. The model should satisfy the relation $f(0) = 1$.

e. The family of models should have a nested structure so that fit can be assessed using a likelihood ratio test.

f. The function should have the property such that it has a slow growth at low doses and also levels off at high doses.

Given the above characterization, Slob (2000) proposes several families of nested models, the most general of which has the form

$$y(d) = \alpha\left[\gamma - (\gamma - 1)\exp\left[-\left(\frac{d}{\beta}\right)^{\delta}\right]\right] \quad \alpha > 0, \beta > 0, \gamma > 0, \delta \geq 1.$$

In addition to the constant model $y = \alpha$, by setting $\delta = 1$, $\gamma = 1$, $\alpha = \gamma = 1$, other models in the hierarchy of nested models are derived. The parameter a in every model represents the background response. Slob (2002) suggests that in practice, models should be fitted consecutively from the simplest to the most general model and statistical criteria such as the likelihood ratio test should be used to decide whether or not the higher step is justified. For illustration, consider once again the data presented in Table 6.2 on the average weight of the pregnant female mice exposed to DEHP. Using the function nml in R, the following four models were fitted to the data using the least squares methodology:

Model 1: $y(d) = \alpha \exp\left(-\frac{d}{\beta}\right)$

Model 2: $y(d) = \alpha \exp\left[-\left(\frac{d}{\beta}\right)^{\delta}\right]$

Model 3: $y(d) = \alpha\left[\gamma - (\gamma - 1)\exp\left(-\frac{d}{\beta}\right)\right]$

Model 4: $y(d) = \alpha\left[\gamma - (\gamma - 1)\exp\left[-\left(\frac{d}{\beta}\right)^{\delta}\right]\right].$

TABLE 6.5

Parameter Estimates for Dose-Response Models

	α	β	γ	δ	SSR
Model 1	23.723	0.24	–	–	31.172
Model 2	22.618	0.226	–	1.953	20.175
Model 3	23.927	0.195	0.09	–	24.342
Model 4	23.927	0.195	0.09	0	24.342

Table 6.5 displays the least square estimate of the parameters for each model along with the associated sum of the squares of residuals (SSRs). Note that if it can be assumed that the continuous response of interest has a normal distribution, the estimates given in Table 6.5 are in fact the maximum likelihood estimates. By a simple examination of the results in Table 6.5, it is evident that Model 2 has the lowest residual sum of squares and it is not justified to use a more complex model. It is interesting to note that for this data set, Model 4 indicates that there is no threshold and exposure at any level could have adverse results.

6.5.1 Mixture of Two Normal Distributions

It is believed in general that if a dose-response model is based on biological principles, it would have a better chance of leading to more realistic results. Razzaghi and Kodell (2000) argue that a problem that frequently occurs in biological experiments with laboratory animals is that some subjects are less susceptible to the treatment than others and that it may frequently happen that some subjects are unaffected by treatment. This so-called nonresponse phenomenon has attracted the attention of many researchers in biological experiments, and several articles have proposed various models that can be used for this problem, see, for example, Boos and Brownie (1986), Salsburg (1986), Conover and Salsburg (1988), and Razzaghi and Nanthakumar (1992, 1994). The data displayed in Table 6.6 pertain to a study of dietary fortification with carbonyl iron. The response variable is the glucose level in C5YSFl black mice. High glucose levels are considered abnormal, and extremely high levels are associated with diabetes. Chen et al. (1996) considered the data set for illustration of a dose-response model based on a single normal distribution. An examination of the data clearly reveals an increase in mean response with increasing dose. But there are also considerable discrepancies in the data at each dose level, there being groups of responses well separated from each other. This suggests that there may be two or more subpopulations, each with greater homogeneity. Thus, in this section we present a method of risk assessment for continuous outcome data due to Razzaghi and Kodell (2000) for situations where the population of animals under study consists of two subpopulations, one consisting of individuals that are less susceptible to a chemical than others and the other, a subpopulation that is considered to be more susceptible to the chemical agent.

TABLE 6.6

Glucose Level in C5YSF1 Black Mice Treated with Carbonyl Iron

Dose	0	15	35	50	100
	62	55	97	67	171
	64	94	96	76	123
	76	100	62	84	118
	59	98	114	104	133
	70	67	71	60	81
	70	106	84	110	97
	76	88	96	142	114
	90	64	87	113	97
	91	82	127	74	71
	80	108	131	124	70
	93	114	101	157	164
	80	124	98	142	130
	62	56	99	67	71
	63	97	64	76	128
	79	101	115	84	116
	56	101	73	104	133
	72	65	84	61	80
	70	106	96	112	97
	75	85	87	142	113
	91	63	125	108	94
	91	87	133	74	69
	79	110	99	121	71
	91	114		154	165
	80	128		143	126
Mean	75.83	92.21	97.23	104.13	113.88
SD	11.36	21.25	20.56	31.44	32.51

As before, let $X_j(d_i); j = 1, 2, ..., n_i$ be the observed value of a random variable $X(d)$ designating the quantitative response variable of interest at dose d. Further, assume that the population of the exposed animals consists of two types of individuals, namely, those animals who have a higher degree of susceptibility to the chemical and those who are relatively less affected by exposure to the toxicant. Then, if we assume that θ and $1 - \theta$ are, respectively, the proportions of the two types of individuals in the population and that the quantitative response of each subpopulation can be modeled by a normal distribution, the distribution of $X(d)$ can be represented by

$$G(x) = P\left[X(d) \leq x\right]$$

$$= \theta \, \Phi\left\{\frac{x - \mu_1(d)}{\sigma_1(d)}\right\} + (1 - \theta) \Phi\left\{\frac{x - \mu_2(d)}{\sigma_2(d)}\right\} \tag{6.21}$$

where $\Phi(.)$ is the cumulative distribution function (cdf) of a standard normal distribution; $\mu_1(d)$, $\mu_2(d)$ are the mean responses; and $\sigma_1(d)$ and $\sigma_2(d)$ are the standard deviations for the two subpopulations. Here, for simplicity, we assume that the standard deviations of the two subpopulations are equal and further that standard deviation is dose-independent. Thus, we assume

$$\sigma_1(d) = \sigma_2(d) = \sigma$$

where σ is the constant common standard deviation. Note that in model (6.21) it is assumed that the proportion of susceptible animals θ is the same in all dose groups. This model is slightly different from that of Boos and Brownie (1991) in the sense that they take θ to be dose dependent and represent the difference in the responses of the two subpopulations by a shift in the mean. However, it is probably more natural to think that the proportion of susceptible animals in the population is a constant. Animals for each dose group (including controls) are selected randomly from the same population, and, apart from random variation, it is more realistic to assume that θ is the same in all dose groups and that differences in responses are reflected in population means. When the response distribution is a single normal, as described before, Kodell and West (1993) suggest that the mean of distribution may be expressed as a polynomial in dose. However, due to restrictions in the number of dose groups in most experimental data, they argue that often a second-degree polynomial should suffice. In their analysis, they found a quadratic equation adequate and satisfactory for describing the mean of the dose-response function. In general, polynomial functions have proved to be quite useful for modeling continuous outcomes, and experience with toxicological data has shown that, in practice, even a quadratic model performs well for risk estimation (Chen et al., 1996; Crump, 1984). Thus, we take

$$\mu_r(d) = \beta_{0r} + \beta_{1r}d + \beta_{2r}d^2 \quad r = 1, 2 \tag{6.22}$$

where we further assume that the mean responses of the two subpopulations at the background rate (control group) are identical, i.e. $\mu_1(0) = \mu_2(0)$ or, equivalently, $\beta_{01} = \beta_{02} = \beta_0$. It can easily be verified that for model (6.14), the mean and variance of $X(d)$ are given by

$$\mu(d) = E[X(d)] = \theta\,\mu_1(d) + (1-\theta)\mu_2(d)$$

$$= \beta_0 + \left[\theta\,\beta_{11} + (1-\theta)\beta_{21}\right]d + \left[\beta_{12} + (1-\theta)\beta_{22}\right]d^2$$

And

$$\sigma^2(d) = \mathrm{Var}(X(d)) = \sigma^2 + \theta(1-\theta)\left[\left(\beta_{11} - \beta_{21}\right)d + \left(\beta_{12} - \beta_{22}\right)d^2\right]^2 \tag{6.23}$$

Thus, even though the variances of the subpopulations are assumed identical, the overall variance of the distribution is dose dependent, being equal to the common variance of the subpopulations at the background and attaining its largest value at a dose level that renders the maximum separation between the mean response functions.

6.5.2 Risk Assessment for the Mixture Dose-Response Model

Once again, define an abnormal adverse effect as a value of $X(d)$ whose magnitude corresponds to values in the tail area of the background distribution, i.e. the distribution of $X(0)$. Thus, abnormal values are those that occur with low probability in unexposed subjects. Clearly, values in either the upper or lower tail may be considered adversely abnormal, but without loss of generality, we only consider values in the lower tail. Hence, a response will be called abnormal if its value is k standard deviations below the mean of the background level distribution, where k is an appropriately chosen constant. Using model (6.14), therefore, the probability of an adverse effect at dose level d is given by

$$P\left[X(d) \le \mu(0) - k\sigma\right] = \theta\, \Phi\left\{\frac{\mu(0) - \mu_1(d)}{\sigma} - k\right\}$$

$$+ (1 - \theta)\Phi\left\{\frac{\mu(0) - \mu_2(d)}{\sigma} - k\right\}.$$

Hence, for a given dose D, if we define the measure of risk as the additional risk, then we have

$$\pi(D) = P\left[X(D) \le \mu(0) - k\sigma\right] - P\left[X(0) \le \mu(0) - k\sigma\right]$$

$$= \theta\left\{\Phi\left(\frac{-\beta_{11}D - \beta_{12}D^2}{\sigma} - \Phi(-k)\right)\right\} \qquad (6.24)$$

$$+ (1 - \theta)\left\{\Phi\left(\frac{-\beta_{21}D - \beta_{22}D^2}{\sigma} - \Phi(-k)\right)\right\}$$

By setting the additional risk $\pi(D)$ at a fixed small value π^*, the maximum likelihood estimates of the unknown parameters of the model are obtained, once unconditionally and once subject to risk restriction

$$\pi(D) = \pi^* \qquad (6.25)$$

and based on the asymptotic distribution of the likelihood ratio statistic, an upper confidence limit for the additional risk can be computed. Accordingly, the $100(1 - \alpha)\%$ upper confidence limit for the excess risk at dose D is the maximum value of $\pi(D)$ that satisfies the risk restriction (6.18) and

$$2\left[\log L_{\max} - \log L(\pi(D))\right] = \chi_{1,2\alpha}^2 \qquad (6.26)$$

where $\log L_{\max}$ represents the maximum value of the log-likelihood function subject to no constraints and $\log L(\pi(D))$ is the maximum value of the log-likelihood subject to constraint (6.23) for a given $\pi(D)$. Equivalently, a lower confidence bound on the BMD for a fixed level of risk is the minimum value of D satisfying (6.25) and (6.26). The quantity $\chi_{1,2\alpha}^2$ is $100(1 - \alpha)\%$ percentile of the χ^2 distribution with 1 degree of freedom (df). Note also that, because of the inherent number of model parameters, in order to ensure identifiability of the model, a bioassay data set with at least five dose groups is required to estimate the parameters.

6.5.3 Maximum Likelihood Estimation

To compute the maximum likelihood estimates in distribution mixtures, Redner and Walker (1984) recommend the use of the Expectation Maximization (EM) algorithm proposed first by Dempster, Laird, and Rubin (1977) for several reasons, including economy of programming and reliable global convergence. The package mixtools in R (Benaglia et al., 2009) provides a thorough comprehensive analysis of mixture models. In fact, many of the algorithms in the mixtools package are based on EM algorithms or on EM-like ideas. Here, we give a description of the algorithm in the context of the current problem. Given the quantitative measurements $x_j(d_i); j = 1, \ldots, n_i, i = 0, \ldots, g$, the log-likelihood function can be written as

$$\log L = \sum_{i=0}^{g} \sum_{j=1}^{n_i} \log \left\{ \theta \, \phi \left[\frac{x_j(d_i) - \mu_1(d_i)}{\sigma} \right] + (1-\theta) \phi \left[\frac{x_j(d_i) - \mu_2(d_i)}{\sigma} \right] \right\} \qquad (6.27)$$

where ϕ (.) denotes the density of a standard normal distribution and $\mu_1(d)$ and $\mu_2(d)$ are given by (6.22). Equation (6.27) has to be maximized in order to derive the maximum likelihood estimates of the parameters. Define the indicator variables Z_{ij} for $j = 1, \ldots, n_i$ and $i = 0, \ldots, g$ as follows:

$$\begin{cases} P(Z_{ij} = 1) = \theta \\ P(Z_{ij} = 0) = 1 - \theta \end{cases}.$$

Then the complete data set is $(X, Z) = \left\{ \left[x_j(d_i), Z_{ij} \right] j = 1, \ldots, n_i, i 0, \ldots, g \right\}$, and, since Z_{ij} are not observed, the EM algorithm can be applied, which proceeds by treating the observations $x = \left\{ x_j(d_i) \right\} j = 1, \ldots, n_i, i = 0, \ldots, g$ as an incomplete data problem. The complete data log-likelihood can now be written in the form

$$\log L_0(\Psi) = N \bar{Z} \log \theta$$

$$+ N(1 - \bar{Z}) \log(1 - \theta) - N \log \sqrt{2\pi} - \left(\frac{N}{2}\right) \log \sigma^2$$

$$- \left(2\sigma^2\right)^{-1} \sum_{i=0}^{g} \sum_{j=1}^{n_i} Z_{ij} \left(x_j(d_i) - \beta_0 - \beta_{11} d - \beta_{21} d^2\right)^2 \qquad (6.28)$$

$$- \left(2\sigma^2\right)^{-1} \sum_{i=0}^{g} \sum_{j=1}^{n_i} (1 - Z_{ij}) \left(x_j(d_i) - \beta_0 - \beta_{12} d - \beta_{22} d^2\right)^2$$

where $\Psi = \left(\theta, \beta_0, \beta_{11}, \beta_{21}, \beta_{12}, \beta_{22}, \sigma^2\right)^T$ is the vector of unknown parameters, $N = \sum_{i=0}^{g} n_i$ and $\bar{Z} = \frac{1}{N} \sum_{i=0}^{g} \sum_{j=1}^{n_i} Z_{ij}$. The EM algorithm now consists of generating from some initial approximation $\Psi^{(0)} = \left(\theta^{(0)}, \beta_0^{(0)}, \beta_{11}^{(0)}, \beta_{21}^{(0)}, \beta_{12}^{(0)}, \beta_{22}^{(0)} \sigma^{2(0)}\right)^T$ a sequence of updated estimates $\Psi^{(m)} = \left(\theta^{(m)}, \beta_0^{(m)}, \beta_{11}^{(m)}, \beta_{21}^{(m)}, \beta_{12}^{(m)}, \beta_{22}^{(m)} \sigma^{2(m)}\right)^T : m = 0, 1, 2, \ldots$ in successive iterations. Each iteration consists of two steps: the E-step and the M-step. The E-step consists of evaluating $Q\left(\Psi, \Psi^{(m)}\right) = E\left(\log L_0(\Psi) \mid X, \Psi^{(m)}\right)$, which is given by (6.20) with Z_{ij} replaced by its conditional expectation,

$$\omega_{ij}^{(m)} = E\left[Z_{ij} / X, \Psi^{(m)}\right]$$

$$= \frac{\theta^{(m)} \phi\left\{x_j(d_i) - \beta_0^{(m)} - \beta_{11}^{(m)} d_i - \beta_{21}^{(m)} d_i^2\right\}}{\theta^{(m)} \phi\left\{x_j(d_i) - \beta_0^{(m)} - \beta_{11}^{(m)} d_i - \beta_{21}^{(m)} d_i^2\right\} + (1 - \theta^{(m)}) \phi\left\{x_j(d_i) - \beta_0^{(m)} - \beta_{12}^{(m)} d_i - \beta_{22}^{(m)} d_i^2\right\}}$$

Note that ω_{ij} and $1 - \omega_{ij}$ can be interpreted as the posterior probabilities that the response $x_j(d_i)$ is, respectively, from an animal that belongs to the more susceptible or less susceptible group. Having computed $Q(\Psi, \Psi^m)$, to obtain the maximum likelihood estimate of Ψ in the unconstrained case, the M-step of the EM algorithm now consists of finding the updated estimate $\Psi^{(m+1)}$ that maximizes $Q(\Psi, \Psi^m)$. Differentiating $Q(\Psi, \Psi^m)$ with respect to the unknown parameters, we obtain a set of seven equations whose solution constitutes the elements of $\Psi^{(m+1)}$. The iteration is continued until a desired level of convergence is reached. See Exercise 5 at the end of this chapter for more details. Once a satisfactory convergence is reached, the unconstrained maximum likelihood $\log L_{\max}$ can be obtained by substitution in (6.27).

6.5.4 Risk Estimation

The upper confidence limit on additional risk at a given dose D is the maximum value of $\pi(D)$ given by (6.24) that satisfies (6.26). Thus, in order to obtain $\log L(\pi(D))$, we need to maximize the log-likelihood (6.28) subject to constraint (6.25). Kim and Taylor (1995) considered an extension of the EM algorithm when the parameters are subject to a linearity constraint. Their methodology would not work here as the constraint (6.25) is nonlinear. Razzaghi and Kodell (2001) proposed an extension of the EM algorithm with nonlinear constraint on the parameters. According to their methodology, to solve this constrained optimization problem, at each iteration, the maximization step is accomplished by using the Lagrange multiplier technique. Hence, the M-step of the algorithm at each iteration consists of finding the maximum of

$$Q^c\left(\Psi, \Psi^{(m)}\right) = N\bar{\omega}^{(m)} \log \theta$$

$$- N\left(1 - \bar{\omega}^{(m)}\right)\log(1 - \theta) + N\log(2\pi)^{-1} - \frac{N}{2}\log\sigma^2$$

$$- \left(2\sigma^2\right)^{-1} \sum_{i=0}^{g} \sum_{j=1}^{n_i} \omega_{ij}^{(m)}\left(x_j(d_i) - \beta_0 - \beta_{11}d - \beta_{21}d^2\right)^2$$

$$- \left(2\sigma^2\right)^{-1} \sum_{i=0}^{g} \sum_{j=1}^{n_i} \left(1 - \omega_{ij}^{(m)}\right)\left(x_j(d_i) - \beta_0 - \beta_{12}d - \beta_{22}d^2\right)^2 \qquad (6.29)$$

$$+ \lambda\left\{\pi(D) - \theta \, \Phi\left(\frac{-\beta_{11}D - \beta_{12}D^2}{\sigma} - k\right)\right.$$

$$\left. + (1 - \theta)\Phi\left(\frac{-\beta_{21}D - \beta_{22}D^2}{\sigma} - k\right) - \Phi(-k)\right\}$$

where λ is the Lagrange multiplier with respect to the unknown parameters. Differentiating (6.29) with respect to $\{\Psi, \lambda\}$ results in a system of eight nonlinear equations in eight unknowns that must be solved numerically using an iterative root-finding algorithm. For the computational details, once again we refer to Razzaghi and Kodell (2000). Note that the roots of the system of equations provide the updated estimates of the parameters. Hence the nonlinear system of equations must be solved at each iteration until a satisfactory convergence is reached, at which time the constrained maximum likelihood $\log L(\pi(D))$ can be obtained.

Next, in order to derive the upper confidence limit on risk, the maximum value of $\pi(D)$ satisfying (6.25) is sought such that (6.26) holds. As mentioned in Section 6.4.2 of this chapter, Kodell and West (1993) describe an algorithm

TABLE 6.7

Benchmark Doses for Balck Mice Exposed to Carbonyl Iron

Additional Risk	BMD	BMDL
0.01	10.893	8.097
0.02	14.561	10.793
0.03	17.087	12.627
0.04	19.101	14.076
0.05	20.821	15.304
0.06	22.35	16.387
0.07	23.745	17.37
0.08	25.042	18.278
0.09	26.267	19.13
0.1	27.436	19.936

whereby the value of the additional risk is incremented by a small amount and the constrained maximization of the likelihood is repeated at each stage of the algorithm. Thus, if the upper confidence limit on the additional risk at the fixed dose D is required, an initial estimate $\hat{\pi}(D)$ is obtained from (6.24) based on the unconstrained maximum likelihood parameter estimates. This initial estimate is incremented by a small amount to get $\hat{\pi}_u(D)$, which is used as the value of π_0 in (6.25) to obtain the maximum constrained log-likelihood $\log L_1(\pi(D))$ subject to $\pi(D) = \hat{\pi}_u^{(0)}(D)$. If (6.26) is not satisfied to a desired accuracy, then $\hat{\pi}_u^{(0)}(D)$ is incremented and the process is repeated. Table 6.7 adapted from Razzaghi and Kodell (2000) gives the values of BMD and BMDL for various risk levels when the above methodology is applied for risk assessment to the data regarding the glucose level of C5YSF1 mice exposed to carbonyl iron represented in Table 6.6.

6.6 A More General Mixture Model

For a cumulative distribution function $G(.)$, a new general class of distributions can be defined as

$$F(x) = \frac{1}{B(\alpha,\beta)} \int_0^{G(x)} \theta^{\alpha-1}(1-\theta)^{\beta-1}\, d\theta \quad \alpha > 0, \beta > 0 \tag{6.30}$$

where $B(\alpha, \beta)$ is the beta function given by

$$B(\alpha,\beta) = \frac{\Gamma(\alpha)\Gamma(\beta)}{\Gamma(\alpha+\beta)}$$

and

$$\Gamma(\alpha) = \int_0^\infty t^{\alpha-1} e^{-t} dt$$

Eugene et al. (2002) used a normal distribution with mean μ and standard deviation σ for $G(.)$ and called the resulting distribution the beta-normal distribution. Thus, from (6.30), the density of the beta-normal distribution is given by

$$f(x) = \frac{1}{\sigma B(\alpha,\beta)} \left\{\Phi\left(\frac{x-\mu}{\sigma}\right)\right\}^{\alpha-1} \left\{1-\Phi\left(\frac{x-\mu}{\sigma}\right)\right\}^{\beta-1} \phi\left(\frac{x-\mu}{\sigma}\right) \quad (6.31)$$

where $\Phi(.)$ and $\phi(.)$ are respectively the cdf and the probability density function (pdf) of a standard normal distribution. Note that the special case $\alpha=\beta=1$ results in the normal distribution. Note also that an equivalent representation of the beta-normal distribution is as follows. Define the random variable X as

$$X = \Phi^{-1}(v)$$

where v is a number between 0 and 1. It is well known that if v is the realized value of a uniform $U(0,1)$ distribution, then X is a standard normal variate. Now, more generally, if v is the realization of a beta random variable with parameters α and β, then the resulting random variable X has a beta-normal distribution, where, as before, in the special case $\alpha=\beta=1$, normality is derived. This latter representation is particularly useful for simulation of beta-normal variables. Eugene et al. (2002) discuss several interesting properties of the beta-normal distribution some of which are worth mentioning here. First, the distribution is symmetric about μ when $\alpha=\beta$, becoming positively skewed when $\alpha>\beta$ and negatively skewed when $\alpha<\beta$. The mean of the distribution is an increasing function of α and a decreasing function of β, while the standard deviation of the distribution is a decreasing function of both α and β. The kurtosis is an increasing function of α for a fixed $\beta<\alpha$ and a decreasing function of β for a fixed $\alpha<\beta$. Eugene et al. (2002) also provide the plots of the beta-normal distribution by varying different variables in the distribution and show that the beta-normal distribution is very flexible and takes a variety of shapes depending on the parameters. In addition, they derive a closed form expression for the first moment of the beta-normal distribution and evaluate it for some specific integer values of α and β. Gupta and Nadarajah (2004) also consider the moments of beta-normal distribution and derive a general expression for moments of the distribution when α and β are integers. The bimodality property of the beta-normal distribution is discussed in Famoye et al. (2004). They show that the distribution becomes bimodal for certain values of parameters α and β. While analytical solutions for values of α and β for which bimodality occurs cannot be obtained, the authors show numerically that generally, bimodality occurs for values of α and β below 0.214. Interestingly, however, the modality of the distribution is independent of the parameters μ and σ. Razzaghi (2009) proposes the use of the beta-normal model (6.24) as a dose-response model for continuous responses. It is argued that although the normal distribution is often used as a dose-response model for quantitative responses, for non-symmetric data, the assumption of normality can lead to erroneous results. Besides, the study Banga et al. (2002) shows that when the normal distribution is used as the dose-response model, the risk estimates are very sensitive to any departure from normality and inferences may not be consistent. The advantage of the class of beta-normal models is that it encompasses a wide range of shapes and the data will determine the most appropriate shape and model. The flexibility and richness of the distribution is an important feature that should help in selecting the correct model and providing an accurate estimate of the risk. As before, we assume that the standard deviation is a constant across all dose levels and that the mean is a quadratic function of the administered dose. Then, for the sample $y_j(d_i); j = 1,\ldots,n_i, i = 0,\ldots,g$ the log-likelihood is given by

$$\log L = N \log \Gamma(\alpha + \beta) - N \log \Gamma(\alpha)$$

$$- N \log \Gamma(\beta) + (\alpha - 1)\sum_{i=0}^{g}\sum_{j=1}^{n_i} \log\left\{\Phi\left[\frac{X_j(d_i) - \beta_0 - \beta_1 d_i - \beta_2 d_i^2}{\sigma}\right]\right\}$$

$$- (\beta - 1)\sum_{i=0}^{g}\sum_{j=1}^{n_i} \log\left\{1 - \Phi\left[\frac{X_j(d_i) - \beta_0 - \beta_1 d_i - \beta_2 d_i^2}{\sigma}\right]\right\} \qquad (6.32)$$

$$- N \log \sigma\sqrt{2\pi} - \frac{1}{2}\sum_{i=0}^{g}\sum_{j=1}^{n_i}\left[\frac{X_j(d_i) - \beta_0 - \beta_1 d_i - \beta_2 d_i^2}{\sigma}\right]^2$$

where $N = \sum_{i=1}^{g} n_i$. Upon setting the derivative of (6.32) with respect to the six parameters α, β, a, b, c, and σ, the following set of six equations is obtained which has to be solved using a numerical root finding algorithm in order to obtain the maximum likelihood estimates of the model parameters.

$$\frac{1}{N}\sum_{i=0}^{g}\sum_{j=0}^{n_i}\log\left\{\Phi\left(z_j(d_i)\right)\right\} + \Psi(\alpha + \beta) - \Psi(\alpha) = 0$$

$$\frac{1}{N}\sum_{i=0}^{g}\sum_{j=0}^{n_i}\log\left\{1 - \Phi\left(z_j(d_i)\right)\right\} + \Psi(\alpha + \beta) - \Psi(\alpha) = 0$$

$$\frac{1}{N}\sum_{i=0}^{g}\sum_{j=1}^{n_i}\left\{(\alpha-1)\frac{\phi(z_j(d_i))}{\Phi(z_j(d_i))}-(\beta-1)\frac{\phi(z_j(d_i))}{1-\Phi(z_j(d_i))}-z_j(d_i)\right\}=0$$

$$\frac{1}{N}\sum_{i=0}^{g}\sum_{j=1}^{n_i}d_i\left\{(\alpha-1)\frac{\phi(z_j(d_i))}{\Phi(z_j(d_i))}-(\beta-1)\frac{\phi(z_j(d_i))}{1-\Phi(z_j(d_i))}-z_j(d_i)\right\}=0$$

$$\frac{1}{N}\sum_{i=0}^{g}\sum_{j=1}^{n_i}d_i^2\left\{(\alpha-1)\frac{\phi(z_j(d_i))}{\Phi(z_j(d_i))}-(\beta-1)\frac{\phi(z_j(d_i))}{1-\Phi(z_j(d_i))}-z_j(d_i)\right\}=0$$

$$\frac{1}{N}\sum_{i=0}^{g}\sum_{j=1}^{n_i}z_j(d_i)\left\{(\alpha-1)\frac{\phi(z_j(d_i))}{\Phi(z_j(d_i))}-(\beta-1)\frac{\phi(z_j(d_i))}{1-\Phi(z_j(d_i))}+1\right\}=1$$

where

$$z_j(d_i)=\frac{x_j(d_i)-\beta_0+\beta_1 d_i+\beta_2 d_i^2}{\sigma}.$$

6.6.1 Risk Assessment with the Beta-Normal Model

Assume now that adverse effects occur on the right tail of the control distribution, although a similar development for experiments that adverse effects occur on the lower end of the control distribution is similarly possible. Hence, if we define an abnormal response as that which falls say $k\sigma$ above the control mean, then the probability of a toxic response at dose d is

$$P(X(d)>\mu(0)+k\sigma)=1-\frac{1}{B(\alpha,\beta)}\int_{0}^{\Phi\left(\frac{\mu(0)+k\sigma-\mu(d)}{\sigma}\right)}\theta^{\alpha-1}(1-\theta)^{\beta-1}d\theta$$

$$=1-\frac{1}{B(\alpha,\beta)}\int_{0}^{\Phi\left(k-\frac{\beta_1 d+\beta_2 d^2}{\sigma}\right)}\theta^{\alpha-1}(1-\theta)^{\beta-1}d\theta.$$

Therefore for a given dose D, if we define the measure of risk $\pi(D)$ as the excess risk over the background risk due to the added fixed dose D, i.e. the additional risk, we have

$$\pi(D) = P\big(X(D) > \mu(0) + k\sigma\big) - P\big(X(0) > \mu(0) + k\sigma\big)$$

$$= \frac{1}{B(\alpha,\beta)}\left\{ \int_0^{\Phi(k)} \theta^{\alpha-1}(1-\theta)^{\beta-1}d\theta - \int_0^{\Phi\left(k - \frac{\beta_1 D + \beta_2 D^2}{\sigma}\right)} \theta^{\alpha-1}(1-\theta)^{\beta-1}d\theta \right\}$$

$$= \int_{\Phi\left(k - \frac{\beta_1 D + \beta_2 D^2}{\sigma}\right)}^{\Phi(k)} \theta^{\alpha-1}(1-\theta)^{\beta-1}d\theta = F[\Phi(k)] - F\left[\Phi\left(k - \frac{\beta_1 D + \beta_2 D^2}{\sigma}\right)\right]$$

(6.33)

where $F(.)$ given by (6.30) is the cdf of the beta distribution. Therefore a point estimate of the additional risk may be obtained by substituting the maximum likelihood estimates of the parameters and evaluating the resulting incomplete beta functions. Note that although the estimate is asymptotically unbiased, it is not so for finite samples. Indeed, if we let

$$\hat{t} = k - \frac{\widehat{\beta_1} D + \widehat{\beta_2} D^2}{\sigma}$$

then an approximate estimate of the bias may be obtained by expanding $F\left[\Phi\left(k - \frac{\widehat{\beta_1} D + \widehat{\beta_2} D^2}{\sigma}\right)\right]$ in the Taylor series about $t = E(\hat{t})$. We have,

$$F[\Phi(\hat{t})] = F[\Phi(t)] + \frac{(\hat{t}-t)}{B(\alpha,\beta)}[\Phi(t)]^{\alpha-1}[1-\Phi(t)]^{\beta-1}\phi(t)$$

$$+ \frac{(\hat{t}-t)^2}{2}\Big\{(\alpha-1)[\Phi(t)]^{\alpha-2}[1-\Phi(t)]^{\beta-1}[\phi(t)]^2$$

$$- (\beta-1)[\Phi(t)]^{\alpha-1}[1-\Phi(t)]^{\beta-2}[\phi(t)]^2$$

$$+ [\Phi(t)]^{\alpha-2}[1-\Phi(t)]^{\beta-1}[\phi'(t)]^2\Big\}$$

which may be used to obtain an estimate of the size of bias.

The beta-normal distribution, as mentioned earlier, is highly flexible and covers a wide range of shapes. Since the data are used to determine the most appropriate shape of the dose-response function, it is believed that the point estimate for the risk at a fixed exposure level or equivalently the point estimate of safe dose for a fixed level of risk *BMD* can naturally be expected to have a high degree of accuracy. We can also use one of the two methods, already discussed, to determine the upper confidence limit on risk. Here, we adopt the method based on asymptotic normality of the maximum likelihood estimates. Since the asymptotic distribution of $\widehat{\beta_1}D + \widehat{\beta_2}D^2$ is normal, assuming $\hat{\sigma}$ is a constant, an approximate upper $100(1 - \alpha)\%$ confidence limit for risk is given by

$$\pi_u(D) = F[\Phi(k)] - F\left[\Phi\left(k - \frac{\hat{\beta}_1 D + \hat{\beta}_2 D^2 + \hat{\tau}(D)z_{1-\alpha}}{\hat{\sigma}}\right)\right] \qquad (6.34)$$

where

$$\tau^2(D) = \mathrm{Var}\left(\hat{\beta}_1 D + \hat{\beta}_2 D^2\right)$$

$$= D^2\mathrm{Var}\left(\hat{\beta}_1\right) + D^4\mathrm{Var}\left(\hat{\beta}_2\right) + 2D^3\,\mathrm{Cov}\left(\hat{\beta}_1, \hat{\beta}_2\right).$$

Another approach that can be adopted for calculating the upper confidence bound on risk is by using the bootstrapping technique. Generating bootstrap samples from the standardized residuals $\dfrac{x_j(d_i) - \hat{\mu}(d_i)}{\sigma}$ the point estimate of risk is calculated for a large number of bootstrap samples and the desired percentile is selected. Through simulation, Chen (1996) shows that all three methods yield very close values (Figure 6.2).

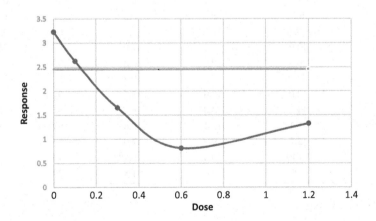

FIGURE 6.2
Quadratic Dose-Response Model and the Abnormal Cut-Off Response for Example 6.2.

Example 6.2

Here, we present the example given in Razzaghi (2009). The data set listed in Table 6.8 is from a 14-day study for finding the dose range of aconiazide conducted at the National Center for Toxicological Research (NCTR). Female Fischer 344 rats were exposed to varying doses of the chemical and change in the body weight was recorded for each rat. Kodell and West (1993) and Chen et al. (1996) used a normal distribution with a quadratic mean function for risk assessment in this data set. Their results show that a normal distribution adequately describes the dose-response fluctuation and suitably estimates the additional risk. Here, we apply the beta-normal distribution which, as described before, includes the normal distribution as a special case and we also use a quadratic mean function. The MODEL procedure in SAS (2003) was used to solve the likelihood equations. The procedure utilizes the Gauss-Newton approach to solve the equations. It was found that convergence was rather rapid and the procedure converged after approximately 10 iterations. Table 6.9 gives the solutions of the likelihood equations along with their respective approximate standard errors obtained from the asymptotic covariance matrix of the estimates. It is interesting to note that the estimates of α and β are both close to one. This confirms the results of Kodell and West (1993) and Chen et al. (1996) as the appropriateness of the normal distribution. Using the methodology based on the asymptotic distribution of maximum likelihood estimates to derive the risk

TABLE 6.8

Weight Gain (g) in Female Fisher 344 Rats Daily Exposed to Aconiazide for 14 days. Data reproduced from Chen et al. (1996)

	Dose (g/kg body weight)				
	0	0.1	0.2	0.5	0.75
	5.7	8.3	9.5	2.9	−8.6
	10.2	12.3	8.1	5.6	0.1
	13.9	6.1	7.0	−3.5	−3.9
	10.3	10.1	7.8	9.5	−4.0
	1.3	6.3	9.3	5.7	−7.3
	12.0	12.0	12.2	4.9	−2.2
	14.0	13.0	6.7	3.8	−5.2
	15.1	13.4	10.6	5.6	−1.0
	12.7	9.9	7.0	4.2	−4.4
Mean	10.40	10.33	8.48	4.43	−4.50
SD	4.25	2.67	1.88	3.29	2.93

TABLE 6.9

Maximum Likelihood Estimate of Model Parameters

Parameter	a	b	c	α	β	σ
Estimate	10.1574	2.18×10^{-4}	-2.88×10^{-5}	0.98626	1.02574	2.986
SE	0.235	0.908	1.743	0.028	0.045	0.098

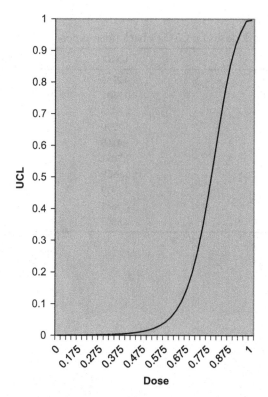

FIGURE 6.3
Upper confidence limit for varying dose levels.

estimates, and taking $k = 3$, so that responses beyond 3σ from the mean of the unexposed animals are considered abnormal, then from the asymptotic covariance matrix of the maximum likelihood estimates, from (6.33) we obtain a point estimate of 0.716×10^{-6} for the additional risk of the lowest dose level of $D = 0.2$ g/kg body weight. Chen et al. (1996) found a point estimate of 9×10^{-4} at this dose level. Similarly, from (6.34) an approximate 95% confidence bound of 0.00074 is obtained on the additional risk at $D = 0.2$ g/kg. Figure 6.3 provides the upper confidence limits for risk versus different values of the additional risk.

We can also use the procedure described in this paper to estimate benchmark doses. Indeed from (6.26) a point estimate of an exposure level corresponding to a low acceptable risk level π^* is the positive root of the quadratic equation

$$\beta_2 D^2 + \beta_1 D + \gamma = 0$$

where

$$\gamma = \sigma \left\{ \Phi^{-1} \left[F^{-1} \left(F \left(\Phi(k) - \pi^* \right) \right) \right] - k \right\}$$

TABLE 6.10

Estimated benchmark doses for C5YSF1 back mice exposed to carbonyl iron

Additional Risk	BMD	95% BMDL
10^{-6}	0.2786	0.0075
2×10^{-6}	0.5601	0.0106
0.03	0.8436	0.0130
0.04	1.1290	0.0150
0.05	1.4165	0.0168
0.06	1.7061	0.0185
0.07	1.9979	0.0199
0.08	2.2919	0.0213
0.09	2.5881	0.0226
0.10	2.8867	0.0238

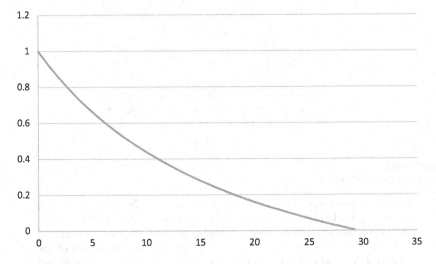

FIGURE 6.4
Saturation Dose-Response Model for Example 6.4.

and a lower confidence bound is similarly obtained using (6.34). Table 6.10 displays the point estimates of the benchmark doses along with the 95% lower confidence limits (BMDLs) on doses corresponding to additional risk levels of 0.01 to 0.10 (Figure 6.4).

6.7 A Semi-Parametric Approach

One can argue that although the distribution of responses in the control group may be normal, the exposure of animals to the test agent can cause a change in the distribution, for example, when only a subset of the animals is affected by the treatment. This argument was used by Razzaghi and Kodell (2001) to apply the mixture model that was introduced earlier in this chapter. Their argument was that during the exposure, some animals are more susceptible to the chemical than others. Thus, the population of the animals can be assumed to consist of two types of individuals, those who are more susceptible and those who are less susceptible. Based on that argument a mixture of two normal distributions was used as the dose-response model. Bosch et al. (1996), however, introduce a semi-parametric approach based on a nonparametric measure of the treatment effect. The dose-response function is characterized as the probability that an animal in the control unexposed group has a higher response than an animal exposed at dose level d; i.e. the risk function at exposure level d is defined as

$$P(d) = P\left[X(0) > X(d)\right] \tag{6.35}$$

where without loss of generality it is assumed that small responses can lead to an abnormal effect. Clearly if responses that fall on the upper tail of the control distribution are abnormal, (6.35) will be adjusted accordingly. Define a vector of indicator variables Y of size $\sum_{i=1}^{g} n_0 n_i$ such that its value in the jrlth position is 1 if the response of the jth animal in the control group exceeds that of the response of the lth animal in the rth dose group, i.e.

$$Y_{jil} = \begin{cases} 1 & \text{if } X_j(0) > X_l(d_i) \ j = 1,\ldots n_0, l = 1,\ldots, n_i, i = 1,\ldots, g \\ 0 \end{cases}$$

Then, from (6.28) since $E(Y_{jrl}) = P(d_i)$, we can model the risk function $P(d)$ as a function of dose using a standard binary regression model such as logic or probit function. We note, however, that the elements of Y are not independent and the jilth element of the diagonal of the covariance matrix is given by

$$V(Y_{jil}) = P(d_i)\left(1 - P(d_i)\right) \tag{6.36}$$

In order to estimate the vector of regression parameters, $\beta^T = \left(\beta_0, \ldots, \beta_p\right)$ we can apply the generalized estimating equation (GEE) technique described in the last chapter. The advantage of this method is that the correlations are treated as parameters that can be estimated. One starts with a working indolence assumption and uses the model to naively estimate the covariance matrix. In this case, the GEE technique leads to solving the equations

$$D^T V^{-1}(Y - P) = 0$$

where P is the $\sum_{i=1}^{g} n_0 n_i$ vector of dose-response whose jilth element is given by $P(d_i)$, V is a diagonal matrix whose jilth diagonal element is given by (6.29) and D is the matrix of the $\sum_{i=1}^{g} n_0 n_i \times p$ matrix of the derivatives $\dfrac{\partial P}{\partial \beta}$. As described in Bosch et al. (1996), solving the above set of equations is equivalent to fitting a binary regression model such as logistic or probit model which is readily available in any standard statistical software. Once the parameter estimates are determined, the naïve estimate of the covariance matrix is found using $\left(D^T V^{-1} D\right)^{-1}$. The method leads to unbiased consistent estimate. For a correct estimate of the covariance matrix, Bosch et al. (1996) suggest an adjustment that can easily be adopted.

Example 6.3

Boose and Brownie (1986) describe an experiment due to Weeks and Collins (1979) in which addiction to morphine in rats is examined. By pressing a lever, rats could obtain morphine by self-injection. After six days of access to morphine, saline was substituted for the morphine sulfate in each dose group, and physical dependence was determined for each rat using a weight loss criterion. The number of lever presses (self-injection rates) on day 6 for the controls and for four concentrations of morphine sulfate, along with the subjective observation of weather the rat was a "responder" or not, are reproduced here in Table 6.11. Good (1979) used the data to illustrate a model for detecting a treatment effect. Based on subjective classification of animals into "responders" and "nonresponders" using the weight loss criterion 50 of the 76 animals in the non-zero dose groups are classified as responders. Boos and Brownie (1986) used the data set to demonstrate that in biological experiments it is not uncommon to exhibit that some animals may be more susceptible to a chemical than others. To convert the count data to continuous responses, they add 1 to each of the count data and then take the logarithm of the results. Razzaghi and Kodell (2001) used the same data set to demonstrate a procedure for an extension of the EM algorithm for constrained optimization. To demonstrate the semi-parametric approach of Bosch et al. (1996) we use the "responder" section of the same data set. Using R, first a Poisson regression model with a quadratic mean dose-response function

$$\log \mu(d) = \beta_0 + \beta_1 d + \beta_2 d^2$$

was fitted to the count data. It was found that the maximum likelihood parameter estimates were

$$\widehat{\beta_0} = 4.923, \ \widehat{\beta_1} = 2.469, \ \widehat{\beta_2} = -3.490.$$

However, after conversion of the data to continuous responses, the GEE method of Bosh et al. (1996) was applied with both the logit and probit link functions, again with a quadratic mean dose-response function. While the

TABLE 6.11

Number of Lever Presses for Self-Injection in Rats

Dose	0	0.032		0.1		0.32		1	
	Count	Count 1	Count 2	Count 1	Count 2	Count 1	Count 2	Count 1	Count 2
	0	82	0	40	0	48	0	8	0
	0	142	16	40	0	64	2	20	
	3	194	23	51	2	103	7	43	
	4	244	49	78	3	111	8	49	
	4	306	101	86	5	123	10	53	
	6	391	126	86	5	146	20	57	
	6	462		96	6	172		62	
	9	536		104	10	190		71	
	9	690		113	13	212		71	
	10	778		123	27	236		77	
	11	918		146	33			97	
	11			159	46				
	13			260	57				
	14			261					
	18			281					
	18			809					
	20			882					
	23								
	26								
	28								
	30								

(Continued)

TABLE 6.11 (CONTINUED)

Number of Lever Presses for Self-Injection in Rats

Dose	0	0.032		0.1		0.32		1	
		Count 1	Count 2	Count 1	Count 2	Count 1	Count 2	Count 1	Count 2
	41								
	50								
	61								
	80								
	94								

Count 1 = Responders, Count 2 = Nonresponders.
Data from Good (1979).

TABLE 6.12

Parameter Estimates for Example 6.3

	β_0	β_1	β_2
Logit	−3.88	0.993	1.331
Probit	−2.05	0.384	0.728

fit for both link functions were very similar, the parameter estimates were different. Table 6.12 gives the parameter estimates for both logit and probit link functions. For the purpose of risk assessment, we use the probit link function as suggested in Bosh et al. (1996).

Suppose that it is desired to estimate the risk at the lowest dosage $D = 0.032$. The point estimate of risk is

$$\hat{P}(D) = \Phi\left(-2.05 + 0.384(0.032) + 0.728(.032)^2\right) = 0.0208$$

and thus the additional risk is

$$\hat{\pi}(D) = \hat{P}(D) - \hat{P}(0) = 0.00062.$$

To find an upper confidence limit for risk, we need the covariance matrix of the parameter estimates. Although Bosh et al. suggest an adjustment to the covariance matrix, in practice the adjustment will not have a significant effect on the risk estimate. The 95% upper confidence limit on risk is given by

$$\widehat{P^u}(D) = \Phi\left\{ \left(1\ D\ D^2\right) \begin{pmatrix} \hat{\beta}_0 \\ \hat{\beta}_1 \\ \hat{\beta}_2 \end{pmatrix} + 1.645\left[\left(1\ D\ D^2\right) C \begin{pmatrix} 1 \\ D \\ D^2 \end{pmatrix} \right]^{1/2} \right\}$$

where C is the estimate covariance matrix given by

$$C = \begin{pmatrix} 0.02383 & -0.13729 & 0.11369 \\ -0.13729 & 1.18497 & -1.05205 \\ 0.11369 & -1.05205 & 0.95011 \end{pmatrix}.$$

Thus we get

$$\widehat{P^u}(D) = \Phi\left\{0.0208 + 1.645(0.128147)\right\} = 0.5915.$$

Accordingly, the benchmark dose is the root of the quadratic equation

$$\hat{\beta}_0 + \hat{\beta}_1 d + \hat{\beta}_2 d^2 = \Phi^{-1}\left[\hat{\pi}_0 + \Phi\left(\hat{\beta}_0\right)\right]$$

where $\hat{\pi}_0$ is a low acceptable risk level. For example, if we set $\hat{\pi}_0 = 0.10$, we find

$$\text{BMD}_{0.10} = 0.5622.$$

To find a lower confidence limit *BMDL* for the benchmark dose, we use an application of the delta method and derive the variance at dose d from

$$\text{Var}(d) = \left(\frac{\partial P(d)}{\partial d}\right)^{-2} \left(\frac{\partial P(d)}{\partial \beta}\right)^{T} C \left(\frac{\partial P(d)}{\partial \beta}\right)^{T}.$$

Here, we have

$$\frac{\partial P(d)}{\partial d} = \left(\beta_1 + \beta_2 d\right)\phi\left(\beta_0 + \beta_1 d + \beta_2 d^2\right)$$

and

$$\left(\frac{\partial P(d)}{\partial \beta}\right) = \begin{pmatrix} 1 \\ d \\ d^2 \end{pmatrix}\phi\left(\beta_0 + \beta_1 d + \beta_2 d^2\right)$$

and thus an estimate of the variance of $BMD_{0.10}$ is

$$\widehat{\text{Var}}(\text{BMD}_{0.10}) = 0.00044803.$$

Therefore

$$\text{BMDL}_{0.10} = \text{BMDL}_{0.10} - 1.645\sqrt{\widehat{\text{Var}}\left(\text{BMD}_{0.10}\right)} = 0.5274.$$

6.8 Dose-Response Modeling and Risk Assessment for Neurotoxic Effect

According to the National Institutes of Health (NIH) web site, neurotoxicity occurs when the exposure to natural or manmade toxic substances (neurotoxicants) alters the normal activity of the nervous system. This can eventually disrupt or even kill neurons, key cells that transmit and process signals in the brain and other parts of the nervous system. Neurotoxicity can result from exposure to substances used in chemotherapy, radiation treatment, drug therapies, and organ transplants, as well as exposure to heavy metals such as lead and mercury, certain foods and food additives, pesticides, industrial and/or cleaning solvents, cosmetics, and some naturally occurring substances. Neurotoxic effects have also been observed from exposure to chlorine, rotten egg gas, ammonia, polyvinyl chloride (PVC), and polychlorinated biphenyls (PCBs). Similar to all non-cancer effects, regulation of neurotoxicants and setting of acceptable exposure levels followed the traditional method of determining the no-observed-adverse-effect-level (NOAEL) and dividing the NOAEL by a series of uncertainty factors. Although, as described in the last chapter, model-based approaches and benchmark dose methodologies have replaced and are preferred in all non-cancer effects, these methods were a little slower for neurotoxicants due to the fact that neurotoxic effects by nature are continuous responses and establishment of the definition of an adverse effect was a little more complex. But, as methodologies were developed for continuous effects, risk assessment approaches for neurotoxicants based on models and benchmark doses became more popular. In this section, we explore how the models that we discussed for continuous outcomes in this chapter are applied to neurotoxic effects.

6.8.1 Defining Adverse Effect for Neurotoxicants

Since the outcome in experiments to assess the neurotoxicity of chemical compounds is continuous in nature, defining an adverse effect has been a subject of controversy. Similar to what was discussed earlier in this chapter, for neurological effects an adverse effect is also defined as an outcome that rarely occurs in unexposed control animals. Thus an abnormal response is defined as an outcome that occurs with a small probability α on the tail distribution of the control animals. Clearly, if small responses are considered abnormal, α is selected on the lower tail, and if large responses are abnormal, it is on the upper tail of the control distribution. The choice of α leads to a cut-off value on the tail beyond which responses are considered to be abnormal. Equivalently, a cut-off value can be selected by choosing a certain number (k) standard deviation below or above the mean of the control distribution depending on whether small or large responses are abnormal. For example, if it can be assumed that the distribution of the control responses is normal, the choice of $\alpha = 0.01$ is equivalent to $k = 2.33$ and a choice of $k = 3$ is equivalent to $\alpha = 0.00135$. Once the cut-off value for abnormality is defined, then for the purpose of risk assessment, one approach may be to dichotomize and classify each response at non-zero doses as either normal or abnormal. However, as discussed earlier, dichotomization is not recommended because of loss of information and the direct approach leads to more informative results.

Now, it is clear that the choice of the tail probability α or equivalently, the choice of the constant k is very crucial and can play a very important role in risk assessment. This choice cannot be arbitrary and must be based on some informed decision. Kodell et al. (1995) consider the problem of how the choice of the constant k can affect the estimated risk. They argue that selection of an appropriate value for the constant k, "provides an opportunity for scientific judgment to enter into the risk assessment process." Although in many experiments for assessing neurotoxicity, the choice of $k=3$ is used, for clinically measured endpoints, three standard deviations below or above the control mean appears to be too extreme and they suggest a tail probability of 1% to 5%, which in the case of normality, leads to choices of 2.33 and 1.64 for k. The paper warns, however, that risk assessors against a small choice of k. As the value of the constant k decreases and therefore the background risk becomes higher, the dose-response relationship across all doses becomes steeper, thus stronger. Thus it is important that care is taken in choosing the value of k. Through an example, it is shown that for most applications, using a small k, or equivalently a relatively large tail area to represent the background risk for adverse effect is not justified. It is recommended that k is selected to be at least 2.

6.8.2 Dose-Response Model

Many of the experiments for testing the neurotoxicity of compounds are the so-called case-control studies in which a group of unexposed individuals are compared with another group exposed at a given dose. However, in experiments with animals and in designing bioassay studies, it is recommended that, if possible, experiments are conducted at several exposure levels. This approach provides information for fitting a suitable dose-response model and thus enabling estimation of risk. Depending on the data, any of the dose-response models for non-quantal continuous data that we discussed earlier in this chapter can be used although the use of the normal and lognormal distributions is most common. To express the mean of the underlying distribution in terms of the exposure level, generally a polynomial function is considered appropriate. In most applications, a polynomial of degree 2 is sufficient, although some authors even used a linear model, see for example Chen et al. (1996). But, Gaylor and Slikker (1992) argue that although a polynomial function may be adequate to express the relationship between the dosage level and the response in the experimental dose range, it provides no assurance of the validity of response estimation at low dosage levels that are experienced by humans. They suggest a saturation model that we introduce in the next section.

6.8.3 A Saturated Model for Neurotoxic Effect

Similar to other toxicological effects, establishment of a mathematical model to describe the relationship between the exposure and the probability of an effect is desirable. As we have seen before, a dose-response model can aid in making informed judgment regarding the effect of a neurotoxicant. The very fact that if the model is based on plausible biological principles, then it will have a much better chance of providing valid estimates hold true for neurotoxicants as well. Slikker and Gaylor (1990) argue that the relationship between the levels of a neurochemical biomarker and the exposed dosage level of a neurotoxicant may be subject to a saturable process. The transformation of the neurotoxic agent to its metabolites may contain steps that must be accounted for in the dose-response models. To this end, rather than using a polynomial, Gaylor and Slikker (1992) suggest modeling the mean response function at exposure level d as a proportion of control mean as

$$\mu(d) = \left(\frac{1 + \beta_0 d}{1 + \beta_1 d} \right) \qquad (6.37)$$

The advantage of this model is that at the background dose of $d = 0$, the level is 100% of the "normal average." Note that at large doses, as $d \to \infty$, the mean response approaches the ratio β_0/β_1. Thus, this ratio may be interpreted as the minimum proportion of the control mean obtained at large doses. Similarly, since

$$\mu(d) = (1 + \beta_0 d) \left[1 - \beta_1 d + \frac{1}{2}(\beta_1 d)^2 - \cdots \right]$$

$$= 1 + (\beta_0 - \beta_1)d + O(d^2),$$

the slope $(\beta_0 - \beta_1)$ may be interpreted as the proportion of the change of the normal average per unit of dose for small doses.

Example 6.4

Table 6.13, adapted from Ricaurte et al. (1987), gives the results of a study on the neurobehavioral effect of methyl-phenyl-tetrahydropyridine (MPTP) in mice. The study consisted of five dose groups with five young mature male C57BL/6J mice. The dopamine concentration in the striata of each mouse was measured one week after four consecutive hourly injections with MPTP. The table gives the sample average concentration

TABLE 6.13

Effect of MPTP on Mice

Dose (mg/kg)	Sample Size	Sample Mean	Sample Variance	Proportion of Controls
0	5	10.4	0.45	1
2.5	5	9.7	0.8	0.9327
5	5	7.7	0.45	0.7403
10	5	5.2	2.45	0.7403
20	5	3.9	0.45	0.375

Data from Ricaurte et al. (1987).

along with the sample variance. A column giving each average as a proportion of the controls is also added to make the data amenable to fitting
the saturated model (6.29). Using the nonlinear modeling function *nlm* in
R, it was found that the model parameter estimates are

$$\hat{\beta}_0 = -0.03394, \quad \hat{\beta}_1 = 0.05051.$$

Thus the initial slope of $(\beta_0 - \beta_1) = -0.084441$ is interpreted as a reduction of about 8.5% in dopamine concentration as the result of 1 mg/kg
MPTP in the low dose region. Also, since $\beta_0/\beta_1 = -0.67206$, at high doses
of MPTP, the saturation limits to about 67% of the mean dopamine level
of control animals. Assume that the dopamine concentration Y has
approximately a lognormal distribution and that the variance σ^2 is dose-
independent. Then, we can get a pooled estimate of the variance as

$$\hat{\sigma}^2 = \exp\left(\frac{0.45 + 0.80 + 0.45 + 2.45 + 0.45}{5}\right)$$

$$= 0.70856.$$

Suppose that abnormal response is defined as an outcome that falls
beyond $k=3$ standard deviation of the control mean. Then, calling this
cut-off response value C,

$$\log C = \log 1 - 3 \times \sqrt{0.70856} = -2.52528$$

$$C = 0.08$$

meaning that with abnormality defined as responses three standard
deviations below the control mean, about 8% or more reduction in dopamine as compared with controls is considered abnormal. This, of course,
is much less than the 67% of the mean control level estimated at high
doses of MPTP. Note that if k is selected to be 2, then $C=0.186$. We will
discuss the role of the choice of k in the next section. Now, at a given dose
D, a point estimate of risk can be derived as

$$P(D) = P\left(Z < \frac{\log C - \log \hat{\mu}(D)}{\hat{\sigma}}\right)$$

$$= P\left(Z < \frac{1}{\hat{\sigma}} \log\left(\frac{C}{\hat{\mu}(D)}\right)\right)$$

and the additional risk is estimated as

$$\hat{\pi}(D) = P(D) - P(0) = P(D) - P(Z < -3).$$

Specifically, if for example $D = 1.0$ *mg/kg*, we have

$$\hat{\mu}(1) = \frac{1 - 0.03394}{1 + .005051} = 0.9196$$

which means about 92% of the control level. A point estimate of risk at this dose level is

$$P(1.0) = P\left(Z < \frac{1}{\sqrt{0.70856}} \log \frac{0.08}{0.9196} \right)$$

$$= P(Z < -2.901) = 0.00186$$

with an additional risk of

$$\hat{\pi}(1) = 0.00186 - 0.00135 = 0.00051$$

Conversely, a point estimate of *BMD* can be derived at a given risk π_0 by finding $\hat{\mu}(d)$ from the equation

$$\Phi^{-1}[\pi_0 + P(0)] = \frac{1}{\hat{\sigma}} \log \left(\frac{C}{\hat{\mu}(d)} \right)$$

and using (6.37) to determine d. Thus *BMD* is the solution of

$$\frac{1 + \hat{\beta}_0 d}{1 + \hat{\beta}_1 d} = C e^{-\hat{\sigma} \Phi^{-1}[\pi_0 + P(0)]}.$$

Thus, for example, if it is desired to find the BMD for an additional risk of say $\pi_0 = 0.0001$, then BMD is the solution of

$$\frac{1 - 0.03394 d}{1 + 0.05051 d} = 0.0065$$

which leads to BMD $= 0.2235$ *mg/kg*.

Clearly, as we have observed before, we seek an upper confidence level for the risk at a given dose and also a lower confidence level for the BMD. But, since we have discussed this topic in several examples in this chapter, we will not pursue that here. The reader is asked to find these values in Exercise 10 at the end of this chapter.

6.8.4 Modeling Time to Response

The complex human nervous system does not control the physiological functions, but it is responsible for controlling many other bodily processes.

The system is very sensitive to exterior perturbations and, in general, has a very limited ability to regenerate. For this reason, it is argued that the scientific community should consider quantification of neurotoxic risk not as a single insult, but rather with regard to an overall impact on human health and studies that consider neurobehavioral effects can be viewed as the first-tier screening. Several studies such as the US Environmental Protection Agency (EPA) superfund study (Moser et al., 1995) consider the so-called functional observational battery (FOB) and measure a multitude of variables in neurobehavioral effects as well as possible adverse effects on neurochemistry, neurophysiology, and neuropathology. The analysis of variance methods with repeated measures has traditionally been applied for the analysis of data from these experiments, and risk assessment has relied on the NOAEL methodology. However, due to the nature of these multi-task experiments, Zhu (2005) argues that the BMD methodology requires estimation of dose for a given low risk size not only at a single instant of time, but rather along a continuum of dose and as well as time between experimental stages. Therefore the dose-response model that is introduced by Zhu (2005) has the following general structure

$$\mu(d,t) = A(t) + f(d,t) \tag{6.38}$$

where it is observed that the mean function μ now also depends on the time t at which a neurobehavioral outcome is observed. The function $A(t)$ represents the behavioral outcomes for the unexposed animals and is only a function of time. Note that in (6.38), it is assumed that the dose effect is additive to the background spontaneous effect. Also, $f(d, t)$ is a smooth function satisfying the initial conditions $f(d, 0) = f(0, t) = 0$. For a specific form of $f(d, t)$, Zhu (2005) suggests

$$f(d,t) = \frac{c_1 dt e^{-\beta t}}{1 + c_2 dt e^{-\beta t}}$$

where c_1, c_2, and β are constants.

Suppose that N animals are exposed and let Y_{ij} $j = 1,\ldots,n_i$, $i = 1,\ldots,N$ be the response of the ith subject at the jth occasion. Then, Y_{ij} is expressed as

$$Y_{ij} = \mu(x_{ij}, \theta_i) + \epsilon_{ij}$$

where the expected value $\mu(x_{ij}, \theta_i)$ is a random effect model with x_{ij} representing the covariates and θ_i is an additive function of the population component and the random component specific to the ith subject. Zhu (2005) applies this model to experimental data on hind limb grip strength of rats exposed to a single dose of triethyl tin (TET) to demonstrate the feasibility of the model in benchmark dose. We will not pursue this mythology any further here and refer the reader to Zhu (2005).

Exercise

1. Using equation (6.4), calculate the Var (benchmark dose, BMD) in Example 6.1 and determine the 95% lower confidence bound *BMDL*.

2. Table 6.6 gives the results of a dietary fortification with carbonyl iron in mice. Use the data set to apply the methodology described in Section 6.4 to fit a lognormal distribution to the data. Determine a point estimate for risk at of 10 dg/kg exposure level and an approximate upper confidence level for risk. Also, determine the 10% benchmark dose and an approximate *BMDL*.

3. Use the data set in Table 6.6 to apply the hierarchical modeling structure of Slob (2006) described in Section 6.6. What model is the most appropriate?

4. Using the log-likelihood (6.11), derive the maximum likelihood estimates of the model parameters β_0, β_1, β_2, and σ^2 for the normal dose-response and a quadratic dose-response function. Generalize by assuming a polynomial of degree l,

$$\mu(d) = \sum_{t=0}^{l} \beta_t d^t$$

for the mean $\mu(d)$ instead of a quadratic function.

5. Using the log-likelihood equation (6.8), substitute the risk constraint (6.10) and derive the maximum likelihood estimates for the parameters β_0, β_1, β_2, and σ^2. Generalize by assuming a polynomial of degree l,

$$\mu(d) = \sum_{t=0}^{l} \beta_t d^t$$

for the mean $\mu(d)$ instead of a quadratic function.

6. Similar to the model described in Section 6.4.2, suppose that the dose-response model is described by the univariate parameter exponential distribution, but rather than the left tail of the distribution of the responses of the control animals, suppose that the direction of adversity is to the right of the control distribution. Thus, assume that responses fall beyond the upper δ percentile of the control distribution. Derive the likelihood equations and show that the probability of an adverse effect at an exposure level d is given by

$$P(d) = P(Y(d) \geq T) = [\delta]^{e^{-\left(\beta_1 d + \beta_2 d^2\right)}}.$$

7. Suppose that the dose-response model is skewed and is expressed by

$$f(y) = \frac{1}{2[\mu(d)]^3} y^2 e^{-y/\mu(d)} \quad y \geq 0,$$

where $\mu(d)$ is the mean function. Assuming that the logarithm of the mean function is a quadratic function of the dose, derive the likelihood equations, and find an expression for the added risk when a response on the lower δth percentile of the control distribution is considered to be abnormal.

8. Verify equation (6.23); i.e. show that when the distribution of the continuous response variable $X(d)$ is a mixture of two normal distributions given by (6.21), then the variance of $X(d)$ is given by (6.23).

9. By differentiating (6.28) with respect to $\Psi = \left(\theta, \beta_0, \beta_{11}, \beta_{21}, \beta_{12}, \beta_{22}, \sigma^2\right)^T$, derive the seven equations for the M-step of the EM algorithm described in Section 3.2 of this chapter. Show that the equations whose solution gives $\Psi^{(m+1)}$, that is the approximation provided at the $(m+1)$th iteration can be expressed as

$$\begin{pmatrix} P^{(m)} & 0 \\ 0^T & 1 \end{pmatrix} \Psi^{(m+1)} = \begin{pmatrix} R^{(m)} \\ s^{(m)} \end{pmatrix}$$

where $P^{(m)}$ is a 6×6 matrix and $R^{(m)}$ is a 6×1 vector that only depend on parameter estimates from the mth step, 0^T is a 6×1 vector of 0's, and $s^{(m)}$ is the estimate of trhe standard deviation derived in the mth step.

Hint: See Razzaghi and Kodell (2000).

10. It is known that the inverse of the Hessian matrix can be assumed to be an approximation for the asymptotic covariance matrix of model parameter estimates. Using the function *nlm* in R, we find that in Example 6.4 the inverse of the Hessian matrix is given by

$$\begin{bmatrix} 0.001356 & 0.009071 \\ 0.009071 & 0.259369 \end{bmatrix}.$$

Using this matrix as the covariance matrix of $\hat{\beta}_0$ and $\hat{\beta}_1$, apply the methodology discussed in this chapter along with the delta method, compute an upper confidence limit for risk at the given dose $D = 1$. Also, for an additional risk of $\pi_0 = 0.0001$, find the lower confidence bound for the BMD.

References

Banga, S. J., Patil, G. P., and Taillie, C. (2000). Sensitivity of normal theory methods to model misspecification in the calculation of upper confidence limits on the risk function for continuous responses. *Environmental and Ecological Statistics*, 7(2), 177–89.

Banga, S., Patil, G. P., and Taillie, C. (2001). Likelihood contour method for calculation of asymptotic upper confidence limits on the risk function for quantitative responses. *Risk Analysis* 21(4), 613–624.

Banga, S., Patil, G. P., and Taillie, C. (2002). Continuous dose-response modeling and risk analysis with gamma and reciprocal gamma distributions. *Environmental and Ecological Statistics* 9(3), 273–293.

Benaglia, T., Chauveau, D., Hunter, D., and Young, D. (2009). mixtools: An R package for analyzing finite mixture models. *Journal of Statistical Software* 32(6), 1–29.

Boos, D. D. and Brownie, C. (1986). Testing for a treatment effect in the presence of nonresponders. *Biometrics* 42, 191–197.

Boos, D. D. and Brownie, C. (1991). Mixture models for continuous data in dose-response studies when some animals are unaffected by treatment. *Biometrics* 47, 1489–1504.

Bosch, R. J., Wypij, D., and Ryan, L. M. (1996). A semiparametric approach to risk assessment for quantitative outcomes. *Risk Analysis* 16(5), 657–665.

Chen, J. J., Kodell, R. L., and Gaylor, D. W. (1996). Risk assessment for nonquantal toxic effects. In *Toxicology and Risk Assessment, Principles, Methods, and Applications*, A. M. Fan and L. W. Chang eds. Marcel Dekker, New York, pp. 503–513.

Conover, W. J. and Salsburg, D. S. (1988). Locally most powerful tests for detecting treatment effects when only a subset of patients can be expected to "respond" to treatment. *Biometrics* 44, 189–196.

Crump, K. S. (1984). A new method for determining allowable daily intakes. *Toxicological Sciences*, 4(5), 854–71.

Dempster, A. P., Laird, N. M., and Rubin, D. B. (1977). Maximum likelihood estimation from incomplete data via the EM algorithm. *Journal of the Royal Statistical Society, Series B (Methodological)*, 39, 1–22.

Eugene, N., Lee, C., and Famoye, F. (2002). Beta-normal distribution and its applications. *Communications in Statistics – Theory Methods* 31(4), 497–512.

Famoye, F., Lee, C., and Eugene, N. (2004). Beta-normal distribution: bimodality properties and applications. *Journal of Modern Applied Statistical Methods* 3, 85–103.

Gaylor, D. W. (1996). Quantalization of continuous data for benchmark dose estimation. *Regulatory Toxicology and Pharmacology* 24(3), 246–250.

Gaylor, D. W. and Slikker, J. W. (1990). Risk assessment for neurotoxic effects. *Neurotoxicology* 11(2), 211–218.

Gaylor, D. and Slikker, Jr, W. (1992). Risk assessment for neurotoxicants. *Neurotoxicolgy*, Eds. Hugh Tilson and Clifford Mitchell, Raven Press, New York, 331–343.

Good, P. I. (1979). Detection of a treatment effect when not all experimental subjects will respond to treatment. *Biometrics* 35, 483–489.

Gupta, A. K. and Nadrajah. S. (2004). On the moments of the beta-normal distribution. *Communications in Statistics – Theory and Methods* 33, 1–13.

Kim, D. K. and Taylor, M. G. (1995). The restricted EM algorithm for maximum likelihood estimation under linear restrictions on the parameters. *Journal of the American Statistical Association* 90, 708–716.

Kodell, R. L. and West, R. W. (1993). Upper confidence limits on excess risk for quantitative responses. *Risk Analysis* 13, 177–182.

Kodell, R. L., Chen, J. J., and Gaylor, D. W. (1995). Neurotoxicity modeling for risk assessment. *Regulatory Toxicology and Pharmacology* 22, 24–29.

Moser, V. C., Cheek, B. M., and MacPhail, R. C. (1995). A multidisciplinary approach to toxicological screening III: Neurobehavioral toxicity. *Journal of Toxicology and Environmental Health* 45, 173–210.

Razzaghi, M. (2009). Beta-normal distribution in dose-response modeling and risk assessment for quantitative responses. *Environmental and Ecological Statistics* 16, 25–36.

Razzaghi, M. and Kodell, R. L. (2000). Risk assessment for quantitative responses using a mixture model. *Biometrics* 56(2), 519–527.

Razzaghi, M. and Kodell, R. L. (2001). An extension of the EM algorithm for optimization of constrained likelihood: An application in toxicology. *Communications in Statistics – Theory and Methods* 30(11), 2317–2327.

Razzaghi, M. and Nanthakumar, A. (1992). On using lehmann alternatives with nonresponders. *Mathematical Biosciences* 80, 69–83.

Razzaghi, M. and Nanthakumar, A. (1994). A distribution free test for detecting a treatment effect in the presence of nonresponders. *Biometrical Journal* 36, 373–384.

Redner, R. A. and Walker, H. F. (1984). Mixture densities, maximum likelihood and EM algorithm. *SIAM Review* 26, 195–239.

Ricaurte, G. A., Irwin, I., Forno, L. S., DeLanny, L. E., Langston, E., and Langston, J. W. (1987). Aging and 1-methyl-4-phenyl-1,2,3,6-tetrahydropyridine-induced degeneration of dopaminergic neurons in the substantia niga. *Brain Research* 403, 43–51.

Salsburg, D. S. (1986). *Statistics for toxicologists*. Marcel Dekker, New York.

SAS. (2003). *Statistical analysis system, version 9.1.* Cary, NC.

Slob, W. (2002). Dose-response modeling of continuous endpoints. *Toxicological Sciences* 66, 298–312.

Slob, W. (2006). Probabilistic dietary exposure assessment taking into account variability in both amount and frequency of consumption. *Food and Chemical Toxicology*, 44(7), 933–51.

Weeks, J. R., & Collins, R. J. (1979). Dose and physical dependence as factors in the self-administration of morphine by rats. *Psychopharmacology*, 65(2), 171–77.

WHO (2008). Guidelines for drinking-water quality, 3rd edition incorporating 1st and 2nd addenda. Vol. 1. Recommendations. Geneva, World Health Organization.

Zhu, Y. (2005). Dose-time-response modeling of longitudinal measurements for neurotoxicity risk assessment. *Environmetrics* 16, 603–617.

7

Developmental Neurotoxicity Modeling and Risk Assessment

7.1 Introduction

Concern over the prenatal exposure of children to chemical substances has had a long history. Developmental neurotoxicity (DNT) is the branch of toxicology that studies the adverse health effects of exogenous agents acting on neurodevelopment. Because human brain development is a delicate process involving many cellular events, the developing fetus is rather more susceptible or perhaps differentially susceptible than the adult nervous system to compounds that can alter the structure and function of the brain. It is known that early exposure to many neurotoxicants can severely damage the developing nervous system. The development of different parts of the central nervous system (CNS) requires the correct number of cells in the proper location for the proper functionality of the system. Cell proliferation, migration, and differentiation are critical in the formation of definite brain structures, and failure of these processes because of the exposure to toxic substances can have a profound deleterious effect on the developing brain. Occurrences of certain disastrous events around the world have become major eye openers for regulatory agencies and risk managers to seriously consider the neurotoxicological consequences of agents on the developing fetus. The thalidomide disaster after the Second World War, leading to the birth of over 10,000 malformed babies in several countries around the world, and the incidence of Minamata disease (Chang and Guo, 1998) in the late 1950s in Minamata Bay, Japan, where, as a result of massive methylmercury exposure, it was observed that the newborns developed symptoms of neurological disorders including mental retardation and cerebral palsy, are just two such examples. The evidence from chemical warfare and weapons of mass destruction is yet another cause for concern in this area. *USAWeekend* published a report by Gosden (1998) who visited Halabja in Iraq ten years after it was bombed with chemical weapons. In that report she states, "On a day I visited the hospital's gynecological ward, no woman had recently delivered a normal baby, but three had just miscarried. The staff spoke of frequent stillbirths and infant deaths; cerebral palsy, Down's syndrome and other mental disabilities." Other similar events such as signs of deficits in IQ and distractibility in children in the early 1970s have produced a clear evidence of maldevelopment of the nervous system in children prenatally exposed to lead. Today, there is a clear evidence that prenatal exposure to many of the environmental chemicals such as heavy metals, pesticides, insecticides, organic solvents, flame retardants, and other organic pollutants can severely damage the developing brain and cause neurodevelopmental damage. Scientists believe that the rise in the incidence of children's neurodevelopment disorders such as lowered IQ, learning disabilities, attention deficit, in particular autism, and attention deficit hyperactivity disorder (ADHD) is associated with the damage caused by this exposure (Smirnova et al., 2014).

According to Giordano and Costa (2012), of over 80,000 chemicals in the market, only about 200 have undergone developmental neurotoxicity testing. In 1998, the US Environmental Protection Agency (US EPA) published testing guidelines for developmental neurotoxicity studies (EPA, 1998). Guidelines for conducting developmental neurotoxicity tests have also been developed by the Organization for Economic Co-operation and Development (OECD, 2003). The latter guidelines were further revised and extended in OECD (2006). Recently (US EPA, 2016), Health Canada and US EPA developed a document on the review and interpretation of the developmental neurotoxicity data to "provide guidance on how to evaluate the quality, the conduct, and the resulting data derived from the behavioral methods employed" in the guidelines. These guidelines generally discuss and point to developmental neurotoxicity testing based on *in vivo* animal (mainly rat) studies. In this chapter, after a brief discussion on the design of developmental neurotoxicity experiments, we present some statistical models for analyzing data from these experiments. We also present some modeling structures that provide ways of estimating the risk. As other non-cancer endpoints, the traditional method for risk assessment for developmental neurotoxicants was based on the calculation of no-observed-adverse-effect-level (NOAEL) as a point of departure and dividing it by a series of safety factors to determine a safe exposure level. But, as in previous chapters, our methodology for risk assessment will be the model based on the estimation of the benchmark dose (BMD) and BMD lower confidence limit (BMDL).

It is to be noted that due to the high cost and certain limitations of current animal test-based guidelines, in recent years there have been many studies on alternative approaches for broader screening of substances with sufficient predictability for human effects. The developmental neurotoxicity scientists believe that current test guidelines are not sufficient to adequately characterize and screen chemical compounds that could potentially harm the developing central nervous system and the brain. These emerging concepts rely on modern technologies pertinent to developmental neurotoxicity testing with the prospect of a paradigm shift in the coming years. The alternative methods are presumably more time and cost efficient and are capable of screening a large number of chemicals. Thus the developmental neurotoxicity scientists are developing new standardized *in vitro* testing methods using cultures derived from human stem cells capable of generating data on the effects of potential neurotoxic substances that represent different stages of the developing brain. During the last few years, the National Toxicology Program (NTP) has evaluated alternative animal models for cell-based analysis of the neurodevelopmental processes. For the purpose of experimentation, species other than rats, such zebrafish and *Caenorhabditis. elegans*, have also been considered. For further information on the justification of the need for new testing of developmental neurotoxicity, we refer to Smirnova et al. (2014) and Bal-Price Bal-Price and Fritsche (2018). Since these new approaches are still in the process of development, statistical models and methods of analyzing data have not been established yet. Indeed, Behl et al. (2019) mention that although the benchmark dose modeling "provides a unified data analysis approach for comparing across datasets, there are still many details to be addressed including guidance for the selection of best-fitting model and/or model averaging, model parameterization, etc." In this chapter therefore, we describe the models that have been used and many of them are still in use for developmental toxicity modeling and risk assessment.

7.2 Bioassay Design of Developmental Neurotoxicity Experiments

Research has shown that there is a high degree of qualitative compatibility between human and animal developmental neurotoxicology (Stanton and Spear, 1990) although there are significant differences between the brain developmental profile of humans and rats which should be taken into account in neurotoxicity studies (Dubovicky et al., 2008). The guidelines developed by the US EPA (1998) and OECD (2011) generally involve experiments with rats consisting of a control and two to three dose groups. Pregnant female dams are randomly assigned to each group. The animals are exposed daily from the time of implantation (gestational day 6) through lactation up to postnatal day 21 to a toxicant during the gestation period. The dams are allowed to spontaneously deliver their pups. Upon term, offspring in each litter are counted, examined for dead pups and external malformations, sexed, and weighed. On or before postnatal day 4, litters are culled by random selection to four males and four females. Equal numbers of male and female pups (usually one or two per task) are assigned to some neurological testing and behavioral examination until adulthood, generally until

TABLE 7.1

An Outline of the EPA Office of Toxic Substances Developmental Neurotoxicity Testing Guidelines

Gestation	Day 0	Sperm and/or plug positive
	Day 6–21	Body weight, OBS*
	Day 6	Begin treatment
Postnatal	Days 1–21	Dams weighed, OBS
	Day 1	Pups counted, weighed, OBS
	Day 4	Litters culled to four males and four females, pups weighed, OBS
	Day 7	Pups weighed, OBS
	Day 13	Motor activity, pups weighed, OBS
	Day 17	Motor activity, pups weighed, OBS
	Day 21	Treatment discontinued at weaning; Motor activity, pups weighed, OBS, neuropathology, brain weight
	Day 22	Auditory startle response, pups weighed, OBS
	Day 45 (±2)	Motor activity, body weight, OBS
	Day 60 (±2)	Motor activity, body weight, active avoidance, auditory startle response, neuropathology, brain weight, OBS
	Days 61–131	Motor activity and exploration, learning and memory, pharmacologic challenge

OBS*: Observations including clinical signs and/or physical landmarks.
Source: Table adapted from Rees et al., (1990) and Slikker and Chang (1998).

postnatal days 120–131. An outline of the systematic approach to developmental neurotoxicity testing developed by the Office of Toxic Substances (OTS) of the EPA is presented in Table 7.1, adapted from Rees et al. (1990).

7.3 Data Analysis

Holson et al. (2008) raise a number of issues in the statistical analysis of data from a DNT study. Here, we briefly mention some of the issues and move on to the next section to discuss the dose-response modeling and risk assessment problems. The testing guidelines point to the use of several procedures (about 15) and in each procedure there are several variables. Some of the variables are examined a number of times and male and female pups are examined separately. Thus a DNT study leads to a large number of dependent variables and in using statistical tests for the analysis, the first problem that is encountered is that a large number of p-values is generated and this may lead to an inflated probability of false-positive results. Thus, to assess significance, it is necessary to control type I error. If α is the acceptable probability of type I error, then the level of significance is adjusted by using $1 - (1 - \alpha)^k$, where k is the number of p-values generated by the tests. Power and sample size are other issues that need to be addressed. Depending on the effect size to be detected, the sample size needs to be adjusted to give the test sufficient power. Using Cohen (1988) effect size criteria, one can approximately determine the required sample size. Holson et al. (2008) compute the power and sample sizes when the probability of type II error is fixed (0.05 and 0.20) for varying effect sizes (medium, large, very large) for a simple one-way analysis of variance (ANOVA) for treatment effects on rats. Thus, for example, to detect a large effect when probability of type II error is fixed at 0.2, one needs 16–20 animals per dose group. In fact experimenters believe that 20 pregnant dams per dose group is perhaps the maximum practically feasible level which provides adequate power for detection of large effects.

The statistical analysis of data in DNT studies generally begins with a test for homogeneity of variance such as the Bartlett's test. Also, some goodness-of-fit criteria are used to determine if it is reasonable to assume normality of the data. This will determine whether or not the ANOVA procedures can be used or some nonparametric methods should be adopted. If necessary, sometimes a transformation is used to make the use of ANOVA criteria justifiable. A simple one-way ANOVA is the most common approach for detecting a treatment effect. Also, when repeated measurements are taken on the same animal, repeated measures ANOVA is used for analysis. This, in turn, entails verification of the sphericity assumption, which is constancy of the variances of the differences across repeated levels of the variance–covariance structure. In addition, Holson et al. (2008) discuss other standard statistical tests such as the analysis of covariance, multiple pairwise comparisons, serial correlations, and interpretation of statistical results.

Suppose that a DNT experiment consists of a control and g non-zero dose levels $0 = d_0 < d_1 < \ldots < d_g$. Assume that m_i pregnant dams are exposed to dose d_i of the chemical and let $\dfrac{n_{ij}}{2}, j = 1, \ldots, m_i$ where n_{ij} is the total number of pups assigned to a task in each litter, be the number of pups from each sex randomly selected for a given task. Let X_{ijkl} be some neurological response of the lth pup of the kth sex of the jth litter of the ith dose group. Then, X_{ijkl} is a quantitative continuous variable and it is the result of a three factor experiment with main factors being dose level, litter, and sex. Hence, the associated linear model can be expressed as

$$X_{ijkl} = \mu + \alpha D_i + \beta L_j + \gamma S_k + \delta(D_i \times S_k) + \varepsilon \tag{7.1}$$

$$i = 0, 1, \ldots, g; \, j = 1, \ldots, m_i; \, k = 1, 2, \, l = 1, \ldots, n_{ij}.$$

Using model (7.1), an analysis of variance can be performed in the usual way (Snedecor and Conchran, 1980) and statistical tests can be conducted for the main effects.

7.4 Dose-Response Modeling

If the analysis of the DNT data indicates a significant dose effect, then the first step for quantitative estimation of risk is the development of an appropriate mathematical dose-response function. Moreover, if the analysis of the data shows that male and female pups respond differently to a developmental neurotoxicant, then separate estimates of risk will be required for each sex. However, if the analysis of variance does not lead to a significant sex effect, then an estimate of risk based on all n_{ij} pups in the jth litter of the ith dose group may be used. Hence, for model development, we need not consider sex as a concomitant variable and therefore without loss of generality we drop the index k from the responses. Also, as in all teratological experiments, in DNT studies, the offspring from the same litter are likely to behave more similarly than offspring from different litters. Because of this so-called intralitter correlation, the usual risk assessment procedures for continuous data such as those described in the last chapter cannot be used in this case and any valid procedure must account for this overdispersion factor. Following Razzaghi and Kodell (2000), suppose that the distribution of X_{ijl} is normal with mean μ_{ij} and variance σ_{ij}^2. As explained before, the normality assumption is reasonable since most continuous type responses encountered in toxicology can adequately be modeled by either a normal or a lognormal distribution. In DNT studies, it is a common practice to report only the sample mean and standard deviation of the observations. Let $\bar{X}_{ij}$ be the sample mean of the n_{ij} responses in the jth litter of the ith dose group. Then

$$P(\bar{X}_{ij} \mid \mu_{ij}) = \left(\frac{n_{ij}}{2\pi\sigma_{ij}^2}\right)^{1/2} \exp\left\{-\frac{n_{ij}}{2\sigma_{ij}^2}\left(\bar{x}_{ij} - \mu_{ij}\right)^2\right\} \tag{7.2}$$

We assume that the standard deviation among litters in the same dose group is the same; i.e. we assume homoscedasticity of responses within each group, that is

$$\sigma_{ij} = \sigma_i \quad j = 1, \ldots, m_i \tag{7.3}$$

To account for the litter effect, suppose that the variation among the means of litters in the same dose group can be characterized by a normal distribution. That is, assume that μ_{ij} has a normal distribution with mean η_i and standard deviation τ_i. Then, the unconditional distribution of $\bar{X}_{ij}$ is derived as

$$P(\bar{X}_{ij}) = \int_{-\infty}^{\infty} \left(\frac{n_{ij}}{2\pi\sigma_i^2\tau_i^2}\right)^{1/2} \exp\left\{-\frac{n_{ij}}{2\sigma_i^2}\left(\bar{x}_{ij} - \mu_{ij}\right)^2 - \frac{1}{2\tau_i^2}\left(\mu_{ij} - \eta_i\right)^2\right\} d\mu_{ij}$$

$$= \left(\frac{n_{ij}}{2\pi\sigma_i^2\tau_i^2}\right)^{1/2} \exp\left\{-\frac{1}{2}\left[\frac{n_{ij}\bar{x}_{ij}^2}{\sigma_i^2} + \frac{\eta_i^2}{\tau_i^2}\right] + \frac{1}{2}\frac{\sigma_i^2\tau_i^2}{\sigma_i^2 + n_{ij}\tau_i^2}\left[\frac{n_{ij}\bar{x}_{ij}}{\sigma_i^2} + \frac{\eta_i}{\tau_i^2}\right]^2\right\}$$

$$\int_{-\infty}^{\infty} \exp\left\{-\frac{\sigma_i^2 + n_{ij}\tau_i^2}{2\sigma_i^2\tau_i^2}\left[\mu_{ij} - \frac{\sigma_i^2\tau_i^2}{\sigma_i^2 + n_{ij}\tau_i^2}\left(\frac{n_{ij}\bar{x}_{ij}}{\sigma_i^2} + \frac{\eta_i}{\tau_i^2}\right)\right]^2\right\} d\mu_{ij}$$

$$= \left[2\pi\left(\frac{\sigma_i^2}{n_{ij}} + \tau_i^2\right)\right]^{-1/2} \exp\left\{-\frac{1}{2\left(\frac{\sigma_i^2}{n_{ij}} + \tau_i^2\right)}(\bar{x}_{ij} - \eta_i)^2\right\}$$

$$(7.4)$$

which shows that the unconditional distribution of $\bar{X}_{ij}$ is normal with mean η_i and variance $\frac{\sigma_i^2}{n_{ij}} + \tau_i^2$. This clearly also means that the unconditional distribution of each individual observation X_{ijl} is normal with mean η_i and variance $\sigma_i^2 + \tau_i^2$.

7.4.1 Intralitter Correlation

We can derive an expression for the intralitter correlation by writing

$$V(\bar{X}_{ij}) = \frac{1}{n_{ij}}V(X_{ijl}) + \frac{1}{n_{ij}^2}\sum\sum_{i \neq j}\text{Cov}(X_{ijl}, X_{ijl'}) \tag{7.5}$$

Now, from (7.4) we have

$$\frac{\sigma_i^2}{n_{ij}} + \tau_i^2 = \frac{1}{n_{ij}}(\sigma_i^2 + \tau_i^2) + \frac{n_{ij}(n_{ij}-1)}{n_{ij}^2}\text{Cov}(X_{ijl}, X_{ijl'})$$

which leads to

$$\rho_i = \text{Cov}(X_{ijl}, X_{ijl'}) = \frac{\tau_i^2}{\sigma_i^2 + \tau_i^2}.$$

Note that the above equation resembles the expression for the intra-class correlation in the one-way random effects model (Snedcor and Cochran, 1980). If we let

$$\theta_i = \frac{\rho_i}{1 - \rho_i} = \frac{\tau_i^2}{\sigma_i^2} \tag{7.6}$$

then θ_i is the overdispersion factor at dose i and $V\left(\bar{X}_{ij}\right)$ can be expressed as

$$V\left(\bar{X}_{ij}\right) = \frac{\sigma_i^2}{n_{ij}}\left(1 + n_{ij}\theta_i\right) \tag{7.7}$$

where the term $n_{ij}\theta_i$ can be interpreted as the overdispersion factor. It is interesting to note the similarity of (7.7) with equation (5.2) from Chapter 5 for the variance structure of the number of responses in developmental toxicity experiments under the beta-binomial model. In that model the variance of the number of responses in the jth litter of the ith dose group is expressed as the variance of the corresponding binomial distribution scaled up by a heterogeneity factor. A similar structure is observed in (7.7). Although it is possible to estimate the overdispersion parameter θ as a dose-dependent parameter, Prentice (1986) recommends using the following model

$$\theta(d) = \frac{\exp\left(\alpha_0 + \alpha_1 d\right) - 1}{2}.$$

However, Razzaghi and Kodell (2004) found that the overdispersion parameter can reasonably be modeled as a function of dose, using a one parameter model

$$\theta(d) = \frac{\exp\left(\alpha(d + \bar{d})\right) - 1}{2} \tag{7.8}$$

Example 7.1

Mohammad and St. Omer (1985, 1986) describe an experiment in which groups of pregnant rats were gavaged with a mixture of herbicides 2,4-dichlorophenoxyacetic acid (2,4-D) and 2,4,5-trichlorophenoxyacetic acid (2,4,5-T) in equal proportions at doses 0, 50, 100, and 125 mg/kg/day. Both of these chemicals are known to be behavioral teratogens in rats. Exposure occurred during the gestation days 6 through 15. Four pups of equal number of males and females from each litter were randomly selected and two males and two females were used for testing physical and behavioral development. Several tests were conducted and results were reported. Here, we use their results on olfactory discrimination response which is the latency in seconds to enter the side of the home nest (for a more thorough description of the test see Mohammad and St. Omer, 1986). Table 7.2 presents dose-specific means and the associated standard error of the mean (SEM) of the test results for postnatal day 9 for each sex. Thus if $\bar{X}_{ij}$ is the sample mean for the jth litter in the ith dose for each sex, then the information provided in Table 7.2 can be interpreted as $\bar{X}_i \pm S_i / \sqrt{m_i}$ for each dose level d_i; $i = 0, \ldots, g$ where

TABLE 7.2

Olfactory Discrimination Response (in Seconds) on Postnatal Day 9 in Rats Exposed to a Mixture of 2,4-Dichlorophenoxyacetic Acid (2,4-D)/2,4,5-Trichlorophenoxyacetic Acid (2,4,5-T)

Dose (mg/kg)	Male		Female	
	Mean	SEM	Mean	SEM
0	51.1	4.8	50.4	6.5
50	55.8	5.2	50.1	5.8
100	63.9	5.4	67.6	6.3
125	74.4	5.6	87.9	7.3

Source: Data from Mohammad and St. Omer (1986).

$$\bar{X}_i = \frac{1}{m_i} \sum_{j=1}^{m_i} \bar{X}_{ij}$$

$$S_i^2 = \frac{1}{m_i - 1} \sum_{j=1}^{m_i} \left(\bar{X}_{ij} - \bar{X}_i\right)^2.$$

Therefore $\bar{X}_i$ and S_i can be interpreted as the estimates for η_i and τ_i, respectively. In the absence of pup-specific or litter-specific data, a small simulation can be performed using the experimental results in Table 7.2 as the basis for generating representative data for illustration. Razzaghi and Kodell (2004) describe the following three-step algorithm for generating a representative data set from summary data for DNT experiments:

Step 1: Generate Litter Means
Generate m_i independent variables from a normal distribution with mean $\bar{X}_i$ and standard deviation S_i. These values can be interpreted as the estimates for the litter means $\hat{\mu}_{ij}, j = 1,\ldots,m_i, i = 0,\ldots,g$.

Step 2: Generate Litter Standard Deviations
For a fixed selected value of the intralitter correlation ρ_i at dose level d_i, calculate the overdispersion factor $\theta_i = \rho_i / (1 - \rho_i)$ for each dose level and use (7.7) to find an estimate of the dose group standard deviation $\hat{\sigma}_i$ from

$$\hat{\sigma}_i = \frac{S_i}{\left(\theta_i + \frac{1}{n_{ij}}\right)^{1/2}}.$$

Step 3: Generate Responses
Generate n_{ij} independent variates from a normal distribution with mean $\hat{\mu}_{ij}$ derived in Step 1 and standard deviation $\hat{\sigma}_i$ derived

in Step 2. These values represent the responses from pups tested for neurological effects in the *j*th litter of the *i*th dose group for each sex.

We emphasize here that the above algorithm leads to a typical representative data set that is not identical to the original experimental data and the generated value is used here for illustration only. Clearly, due to sampling variability, the results will vary for each data set that is generated. However, the variation is not expected to influence the results of our analysis significantly. For the purpose of analysis, typical values of $n_{ij}=2$ pups from each litter with equal number of dams $m_i=16$ per dose group may be applied. To induce the intralitter correlations, Razzaghi and Kodell (2004) suggest the value of $\alpha=0.003$ in (7.8). This value, in turn, leads to the values of 0.100, 0.175, 0.250, and 0.285 for the overdispersion parameter at 0, 50, 100, and 125 dose levels, respectively. Generally, in developmental toxicity experiments, the value of the overdispersion parameter is expected to rise with increasing dose and usually ranges between 5% and 40% (Kodell et al., 1991).

Tables 7.3 and 7.4, adapted from Razzaghi and Kodell (2004) respectively, give the generated representative sample observations for male pups and female pups. Using the linear model 7.1, it was found that dose, sex, and their interaction were significant effects. Table 7.5 displays the associated *R* output. Running the analysis of variance for each sex separately, it was found that dose effect was significant for female pups and not the male pups.

7.4.2 Maximum Likelihood Estimation

As we did in the last chapter for neurotoxic responses, in order to model the mean function at each dose level, we use a polynomial. As explained before, in practice, a polynomial of second degree will suffice. Thus,

$$\eta(d) = \sum_{r=0}^{p} \beta_r d^r \qquad (7.9)$$

Then, using (7.4) and (7.7) the log-likelihood is given by

$$L = \sum_{i=0}^{g} \sum_{j=1}^{m_i} \left\{ -\frac{1}{2} \log\left[2\pi \frac{\sigma_i^2}{n_{ij}} \left(1 + n_{ij}\,\theta(d_i)\right) \right] - \frac{\left(\bar{X}_{ij} - \sum_{r=0}^{p} \beta_r\, d^r\right)^2}{\dfrac{\sigma_i^2}{n_{ij}}\left(1 + n_{ij}\,\theta(d_i)\right)} \right\} \qquad (7.10)$$

Substituting (7.8) for $\theta(d)$ and setting the derivatives of (7.10) with respect to all of the parameters $\alpha, \beta_0, \ldots, \beta_p, \sigma_0^2, \ldots, \sigma_g^2$, the following set of equations is derived:

TABLE 7.3

Individual Female Pup Olfactory Responses

Litter	Dose (mg/kg/day)							
	0		50		100		125	
	Pup 1	Pup 2	Pup 1	Pup 2	Pup 1	Pup 2	Pup 1	Pup 2
1	48.24	6.25	61.13	92.84	55.55	203.26	82.23	7.64
2	99.41	67.30	56.65	87.07	93.89	103.51	74.36	71.21
3	31.41	102.71	11.81	78.21	95.01	35.62	85.41	119.06
4	76.81	33.12	78.72	20.08	24.13	1.06	43.61	71.25
5	51.47	62.41	61.43	84.64	51.21	43.40	132.80	118.05
6	42.86	28.28	24.94	5.00	124.82	111.11	57.09	1.47
7	44.63	46.98	67.74	87.87	41.57	94.74	70.88	96.86
8	95.81	53.38	1.60	8.04	24.88	36.46	129.69	103.75
9	18.08	0.88	25.07	61.10	117.18	50.72	72.14	57.47
10	16.13	52.20	48.62	50.18	74.49	11.61	12.89	76.77
11	24.98	88.51	22.99	34.88	66.00	51.08	175.53	235.85
12	16.08	13.91	65.55	42.71	45.66	54.49	64.41	184.09
13	87.55	20.59	87.74	17.53	41.03	82.26	98.81	113.28
14	5.08	78.03	129.74	85.86	100.75	88.54	87.43	71.26
15	41.83	110.11	10.41	19.11	126.42	47.44	99.25	28.31
16	60.92	49.77	0.37	68.89	25.53	90.29	99.26	103.45
Mean	49.2		50.0		69.2		88.9	

Source: Data from Razzaghi and Kodell (2004).

TABLE 7.4

Individual Male Pup Olfactory Responses

Litter	Dose (mg/kg/day)							
	0		50		100		125	
	Pup 1	Pup 2	Pup 1	Pup 2	Pup 1	Pup 2	Pup 1	Pup 2
1	84.28	67.16	74.90	80.03	92.87	74.24	85.72	68.22
2	16.68	21.94	42.72	53.02	101.23	76.06	121.25	116.34
3	36.47	49.08	30.71	93.32	89.59	28.71	94.37	95.79
4	70.12	54.13	32.71	17.55	37.62	10.55	74.12	76.88
5	84.53	70.72	48.64	77.74	96.54	124.20	19.07	10.76
6	62.81	12.38	87.62	66.04	89.91	63.38	117.17	16.14
7	93.66	18.05	84.66	60.58	72.04	35.71	51.75	13.05
8	74.75	54.87	48.96	59.69	25.17	62.36	43.66	9.71
9	43.72	35.31	41.42	22.11	29.93	52.97	67.95	21.97
10	73.52	53.26	0.16	35.19	7.57	68.36	90.44	27.26
11	54.31	93.08	43.84	60.46	59.13	81.61	45.44	4.13
12	40.74	47.49	102.23	43.19	118.22	181.71	31.72	0.65
13	62.97	48.26	63.94	97.10	29.99	80.17	74.34	3.61
14	8.33	2.68	48.89	72.28	80.99	56.21	80.25	42.30
15	84.19	82.39	31.57	39.70	34.04	37.83	43.92	17.16
16	4.10	9.27	63.27	41.57	31.86	20.74	62.76	16.64
Mean	50.5		55.4		64.1		74.7	

Source: Data from Razzaghi and Kodell (2004).

TABLE 7.5

R Output for Linear Model 7.1

	Df	Sum Sq	Mean Sq	F value	Pr(>F)	
Dose	3	19536	6512	5.195	0.00172	**
Litter	15	28936	1929	1.539	0.09260	
Sex	1	5224	5224	4.167	0.04234	*
Dose:Sex	3	18187	6062	4.837	0.00276	**
Residuals	233	292056	1253			

$$\frac{\partial L}{\partial \alpha} = 0 \rightarrow \sum_{i=0}^{g}\sum_{j=1}^{m_i} \frac{n_{ij}\left(d_i + \bar{d}\right)\left[\theta(d_i) + 1/2\right]}{1 + n_{ij}\theta(d_i)}\left\{1 - \frac{n_{ij}\left[\left(\bar{X}_{ij} - \sum_{r=0}^{p}\beta_r d^r\right)^2\right]}{\sigma_i^2\left(1 + n_{ij}\theta(d_i)\right)}\right\}$$

$$= 0 \quad s = 0,1$$

$$\frac{\partial L}{\partial \beta_s} = 0 \rightarrow \sum_{i=0}^{g}\sum_{j=1}^{m_i}\left[\frac{d_i^s\left(\bar{X}_{ij} - \sum_{r=0}^{p}\beta_r d^r\right)}{\sigma_i^2\left(1 + n_{ij}\theta(d_i)\right)}\right] = 0 \quad s = 0,\ldots,g$$

$$\frac{\partial l}{\partial \sigma_s^2} = 0 \rightarrow \sigma_s^2 = \frac{1}{m_s}\sum_{j=1}^{m_s}\frac{n_{ij}\left[\left(\bar{X}_{sj} - \sum_{r=0}^{p}\beta_r d^r\right)^2\right]}{1 + n_{sj}\theta(d_i)} \quad s = 0,\ldots,g$$

The equations can be solved using a standard root finding procedure such as the Newton-Raphson method. If it can further be assumed that the variance σ_i^2 is the same for all dose levels, that is, if $\sigma_i^2 = \sigma^2; i = 0,\ldots,g$, then the last set of likelihood equations above collapses into one equation as follows:

$$\sigma^2 = \frac{1}{N}\sum_{i=0}^{g}\sum_{j=1}^{m_i}\frac{n_{ij}\left[\left(\bar{X}_{ij} - \sum_{r=0}^{p}\beta_r d^r\right)^2\right]}{1 + n_{ij}\theta(d_i)}$$

where $N = \sum_{i=0}^{g} m_i$. The assumption of homogeneity of variance at all dose levels is not unreasonable and substantially reduces the number of equations. In fact, Ryan et al. (1991) and Catalano and Ryan (1992) consider the problem of modeling continuous outcomes in developmental toxicity experiments and find comparable variation among mean responses from different groups.

TABLE 7.6

Maximum Likelihood Estimate (MLE) of the Model Parameters

Parameter	β_0	β_1	β_2	α	σ^2
MLE (female pups)	49.309	−0.1992	0.0041	0.0103	402.412
MLE (male pups)	50.698	0.0056	0.0014	0.0031	796.938

Example 7.2

In Example 7.1, we described an experiment by Mohammad and St. Omer (1985, 1986) about the neurodevelopmental effects of a mixture of herbicides 2,4-D and 2,4,5-T in rats. The results of olfactory discrimination response on postnatal day 9 were used for detecting the treatment effect. Since only the dose level summary data was available, litter means were generated based on a representative sample (see Tables 7.4 and 7.5). Razzaghi and Kodell (2004) use the litter means for the male and female pups separately to fit the hierarchical normal structure described in this section. To model the mean function $\eta(d)$, a polynomial of degree 2, i.e. $p=2$ in (7.9) is used. As discussed in the previous chapter, a quadratic equation is considered quite adequate for modeling the mean function. For simplicity, it was further assumed that the variance is dose independent and that variances across all dose levels were the same; i.e.

$$\sigma_i^2 = \sigma^2; i = 0,\dots,g.$$

The assumption of homogeneity of variance is not unreasonable as comparable variation among mean responses from different dose groups have been found by Ryan et al. (1991) and Catalano and Ryan (1994) in studies of continuous responses from developmental toxicity experiments. Table 7.6 displays the maximum likelihood estimate of the model parameters.

7.5 Risk Assessment

Research in the field of behavioral teratology and developmental neurotoxicology dates back to the 1960s when Werboff and Gottlieb (1963) reviewed the work on postnatal consequences of prenatal exposure to x-irradiation. However, it was not until early 1970s that the potential harm of prenatal exposure to developmental neurotoxicants in humans was formally acknowledged. The traditional method for setting an acceptable exposure level for developmental neurotoxic effects, similar to all non-cancer endpoints, was through the application of uncertainty factors and the NOAEL. A NOAEL, which is the experimental dose level immediately below the lowest dose that produces a statistically or biologically significant increase in the rate of adverse effects over controls (Barnes and Dourson, 1989), is first determined. The NOAEL is then divided by a series of safety factors to determine a reference dose for humans. In fact Tilson (2000) recommends the same method for developmental neurotoxicants. But, as discussed before, there is clear evidence that methods based on the NOAEL approach are not fully reliable. Gaylor (1994) shows that the NOAEL-based method is ill-defined, somewhat subjective, highly dependent on the dose spacing, and does not reward better experimentation. Model-based approaches have proved to be more reliable and have replaced the traditional methods.

As in all toxicological experiments with continuous responses, for developmental neurotoxicity outcomes, one of the difficulties in risk assessment for neurotoxic effects is that the endpoints of interest are quantitative measures and there is not a clear definition of the degree of change that is considered "adverse." In such cases, as we saw in the last chapter, one approach is to define an abnormal response as that whose quantitative value corresponds to the tail area of the distribution of responses corresponding to the control animals. This approach was first proposed by Gaylor and Slikker (1990) and has since been adopted in many studies; see for example, Chen et al. (1996) and Razzaghi and Kodell (2000). Let $X(d)$ the random variable designating a neurological response to a developmental neurotoxicant at dose d. Assume that the distribution of $X(d)$ is normal with mean $\mu(d)$ and variance $\sigma^2(d)$. Then, if it can be assumed that fluctuations among the responses from different litters in the same dose group are reflected in the mean response and that the fluctuations can be represented by a normal distribution with mean (d) and variance $\tau^2(d)$, then from (7.4) we can say that the unconditional distribution of $X(d)$ is normal with mean $\eta(d)$ and variance $\sigma^2(d)+\tau^2(d)$. Thus, the risk function $R(d)$, that is the probability of an abnormal response at exposure level d, defined as responses that fall k standard deviations below the mean of the control responses, can be expressed as

$$R(d) = P\left[X(d) \leq \eta(0) - k\left(\sigma^2(0)+\tau^2(0)\right)^{\frac{1}{2}} \right]$$

$$= \Phi\left\{ \frac{\left(\eta(0)-\eta(d)\right)-k\left(\sigma^2(0)+\tau^2(0)\right)^{\frac{1}{2}}}{\left(\sigma^2(d)+\tau^2(d)\right)^{\frac{1}{2}}} \right\}$$

and using (7.6), we have

$$R(d) = \Phi\left\{ \frac{\left(\eta(0)-\eta(d)\right)-k\sigma(0)\left(1+\theta(0)\right)^{1/2}}{\sigma(d)\left(1+\theta(d)\right)^{1/2}} \right\} \tag{7.11}$$

where k is an appropriately chosen constant (usually 2.33 or 3.0) that yields a specific low response rate in control animals and $\theta(d)$ represents overdispersion at dose d. In the last chapter, we presented a discussion for the choice of the constant k and how the choice may affect the risk estimates. Also, in (7.11), it is assumed that an adverse effect occurs for low responses. Clearly, if an abnormal response is a response that is on the upper tail of the control distribution, a similar expression may be derived. If we model the mean function $\eta(d)$ as a polynomial in d, as in (7.9) and if we define the measure of risk $\pi(D)$ at a given exposure level D as the additional risk, that is the excess risk above the ambient background exposure, then

$$\pi(D) = R(d) - R(0) = \Phi\left\{\frac{-\sum_{r=1}^{p}\beta_r D^r - k\sigma(0)\left(1+\theta(0)\right)^{\frac{1}{2}}}{\sigma(D)\left(1+\theta(D)\right)^{\frac{1}{2}}}\right\} - \Phi(-k) \quad (7.12)$$

Thus a point of the additional risk $\hat{\pi}(D)$ may be found by substituting the maximum likelihood estimates of the model parameters as discussed in the last section in (7.12).

To derive an upper confidence limit for the added risk, one approach is to apply the asymptotic properties of the maximum likelihood estimates and the delta method as discussed previously. A more common approach, however, is to utilize the asymptotic property of the likelihood ratio statistics. Accordingly, if $L_{\max}$ is the maximum value of the likelihood function, then the $100(1 - \alpha)\%$ upper confidence limit for the excess risk at dose D is the maximum value of $\hat{\pi}(D)$ that satisfies

$$2\left\{\log L_{\max} - \log L\left[\pi(d)\right]\right\} = \chi^2_{1,2\alpha} \quad (7.13)$$

where $L[\pi(d)]$ is the maximum of the likelihood function subject to the risk restriction $\pi(D)$ given in (7.12). In order to impose this restriction, the simplest approach would be to solve (7.12) for one of the parameters and substitute in the likelihood function. Specifically, if we assume that the mean is a quadratic function of dose, we can solve (7.12) for either β_1 or β_2 and substitute in the likelihood function (7.10) and the resulting function is maximized with respect to the remaining parameters. For example, if we solve (7.12) with respect to β_2, we get

$$\beta_2 = \frac{-\sigma(D)\left(1+\theta(D)\right)^{\frac{1}{2}}}{D^2}\left\{\Phi^{-1}\left[\pi(D)+\Phi(-k)\right]+k\left[\frac{1+\theta(0)}{1+\theta(D)}\right]^{1/2}\right\} - \frac{\beta_1}{D}$$

which must be substituted in (7.10) in order to obtain $L[\pi(d)]$.

In order to find the maximum value of $\pi(D)$ that satisfies (7.13), an algorithm by Kodell and West (1993) was described in the last chapter whereby

TABLE 7.7

Benchmark Doses for Olfactory Discrimination Response

	Male		Female	
	BMD 0.05	BMDL 0.05	BMD 0.05	BMDL 0.05
Max. likelihood	73.7	25.3	55.8	23.7
Kavlock et al.	66.3	26.5	81.1	31.3
NOAEL		100		50

the value of excess risk from the unrestricted maximum likelihood $\hat{\pi}(D)$ is incremented by a small amount and log $L[\pi(d)]$ is calculated based on this incremented value. If (7.13) is satisfied, the incremented value is the desired upper confidence of risk. Otherwise the excess risk is incremented again and the process is repeated until equality in (7.13) is achieved. Clearly, the same algorithm can be applied here.

Example 7.3

Continuing from Examples 7.1 and 7.2 regarding the olfactory discrimination response data on postnatal day 9, Table 7.7 gives the $BMD_{0.05}$ and the $BMDL_{0.05}$ for both male and female pups, where for comparison the corresponding values using a procedure by Kavlock et al. (1995) and the NOAEL values are also given. The Kavlock et al. (1995) procedure uses the so-called no-statistical-significance-of-trend approach, whereby linear contrasts in dose are tested using ANOVA using F-test with the mean square for litters within doses as the error term. Since the approach has not found wide application compared to the benchmark dose approach, we will not discuss it here and refer to Kavlock et al. (1995) for further details.

7.6 Modeling DNT Responses with Repeated Measures

In practice, often several postnatal observations are made on a variable on each pup. For example, in Table 7.1, summarizing the design of a typical DNT study, observation on the motor skills of the pup is called for on postnatal days 13, 17, 45, 60, and beyond. In this section, we develop a modeling structure that accounts for this multiplicity of measurements. Suppose that a total of k sequential repeated measurements of a variable X are taken at times t_1, ..., t_k. Let $X_{ijl}(t_r)$, $r = 1, \ldots, k$ be the rth measurement on the lth pup in the jth litter of the ith dose group for $l = 1, \ldots, n_{ij}$; $j = 1, \ldots, m_i$; $i = 0, \ldots, g$, where, in practice, n_{ij} is usually 2 or 4. Let $\bar{X}_{ij}(t_r)$ be the mean of measurements on the lth pup, that is, let

$$\bar{X}_{ij}(t_r) = \frac{1}{n_{ij}} \sum_{l=1}^{n_{ij}} X_{ijl}(t_r).$$

Assume that the distribution of $X_{ijl} = \left[X_{ijl}(t_1),\ldots,X_{ijl}(t_k)\right]^T$ be multivariate normal with mean $\mu_{ij} = \left[\mu_{ij}(t_1),\ldots,\mu_{ij}(t_k)\right]^T$ and $k \times k$ covariance matrix V_{ij}. Then, conditional on μ_{ij}, the distribution of $\bar{X}_{ij} = \left[\bar{X}_{ij}(t_r),\ldots,\bar{X}_{ij}(t_r)\right]^T$ is multivariate normal with mean μ_{ij} and covariance matrix $\dfrac{1}{n_{ij}}V_{ij}$.

Let $\Sigma_{ij} = \dfrac{1}{n_{ij}}V_{ij}$ and suppose now that variations among n_{ij} responses in the jth litter of the ith dose group can be characterized by a normal distribution; i.e. assume that $\mu_{ij}(t_r)$ has a normal distribution with mean $\eta_i(t_r)$ and variance $\tau_i^2(t_r)$. We are further going to assume that the litter means $\mu_{ij}(t_1),\ldots,\mu_{ij}(t_k)$ are independent and that, for simplicity, variances are homogeneous, i.e. $\tau_i^2(t_r) = \tau^2$, although these assumptions may not always be true. Then μ_{ij} has a multivariate normal distribution with mean $\eta_i = \left[\eta_i(t_1),\ldots,\eta_i(t_k)\right]^T$ and covariance matrix $\tau^2 I$ where I is the $k \times k$ identity matrix. Now, the unconditional distribution of $\bar{X}_{ij}$ is given by the k fold integral

$$P\left(\bar{X}_{ij}\right) = \int_{-\infty}^{\infty} \cdots \int_{-\infty}^{\infty} (2\pi)^{-\frac{k}{2}} \left|\Sigma_{ij}\right|^{-\frac{1}{2}} \exp\left[-\frac{1}{2}\left(\bar{X}_{ij} - \mu_{ij}\right)^T \Sigma_{ij}^{-1}\left(\bar{X}_{ij} - \mu_{ij}\right)\right]$$

$$\cdot (2\pi)^{-\frac{k}{2}}\tau^{-\frac{k}{2}} \exp\left[-\frac{1}{2\tau^2}\left(\mu_{ij} - \eta_i\right)^T\left(\mu_{ij} - \eta_i\right)\right] d\mu_{ij}$$

$$= \int_{-\infty}^{\infty} (2\pi)^{-k}\tau^{-k/2}\left|\Sigma_{ij}\right|^{-\frac{1}{2}} \exp\left\{-\frac{1}{2}\left[\bar{X}_{ij}^T\Sigma_{ij}^{-1}\bar{X}_{ij} - 2\bar{X}_{ij}^T\Sigma_{ij}^{-1}\mu_{ij} - \mu_{ij}^T\Sigma_{ij}^{-1}\mu_{ij}\right.\right.$$

$$\left.\left. + \tau^{-2}\left(\mu_{ij}^T\mu_{ij} - 2\mu_{ij}^T\eta_i + \eta_i^T\eta_i\right)\right]\right\} d\mu_{ij}$$

$$= (2\pi)^{-k}\left|\Sigma_{ij}\right|^{-\frac{1}{2}}\tau^{-k/2}\exp\left\{-\frac{1}{2}\left(\bar{X}_{ij}^T\Sigma_{ij}^{-1}\bar{X}_{ij} + \frac{1}{\tau^2}\eta_i^T\eta_i\right)\right\}$$

$$\int_{-\infty}^{\infty} \exp\left\{-\frac{1}{2}\left[\mu_{ij}^T\left(\Sigma_{ij}^{-1} + \tau^{-2}I\right)\mu_{ij} - 2\left(\bar{X}_{ij}^T\Sigma_{ij}^{-1} + \frac{1}{\tau^2}\eta_i^T\right)\mu_{ij}\right]\right\} d\mu_{ij}$$

$$(7.14)$$

Now, using a result from matrix theory (see for example lemma 10 in Searle, 1971) we can have

$$P\left(\bar{X}_{ij}\right) = (2\pi)^{-k}\left|\Sigma_{ij}\right|^{-\frac{1}{2}}\tau^{-k/2}\exp\left\{-\frac{1}{2}\left(\bar{X}_{ij}^T\Sigma_{ij}^{-1}\bar{X}_{ij} + \frac{1}{\tau^2}\eta_i^T\eta_i\right)\right\}$$

$$\cdot (2\pi)^{-k/2} \left| \Sigma_{ij} + \tau^{-2}I \right|^{-\frac{1}{2}}$$

$$\exp\left\{ -\frac{1}{2} \left(\bar{X}_{ij}^{T} \Sigma_{ij}^{-1} + \frac{1}{\tau^2} \eta_i^{T} \right) \left(\Sigma_{ij}^{-1} + \tau^{-2}I \right)^{-1} \left(\bar{X}_{ij}^{T} \Sigma_{ij}^{-1} + \frac{1}{\tau^2} \eta_i^{T} \right)^{T} \right\}$$

(7.15)

Note that

$$\left(\Sigma_{ij}^{-1} + \tau^{-2}I \right) = \Sigma_{ij}^{-1} \tau^{-2} \left(\Sigma_{ij} + \tau^2 I \right)$$

and so

$$\left(\Sigma_{ij}^{-1} + \tau^{-2}I \right)^{-1} = \Sigma_{ij} \tau^2 \left(\Sigma_{ij} + \tau^2 I \right)^{-1}$$

And that

$$\left| \Sigma_{ij} + \tau^{-2}I \right|^{-\frac{1}{2}} = \tau^{k/2} \left| \Sigma_{ij} \right|^{\frac{1}{2}} \left| \Sigma_{ij} + \tau^2 I \right|^{-\frac{1}{2}}$$

(7.16)

Also,

$$\bar{X}_{ij}^{T} \Sigma_{ij}^{-1} \bar{X}_{ij} - \bar{X}_{ij}^{T} \Sigma_{ij}^{-1} \left(\Sigma_{ij}^{-1} + \tau^{-2}I \right)^{-1} \Sigma_{ij}^{-1} \bar{X}_{ij}$$

$$= \bar{X}_{ij}^{T} \Sigma_{ij}^{-1} \left(\Sigma_{ij} + \tau^2 I \right) \left(\Sigma_{ij} + \tau^2 I \right)^{-1} \bar{X}_{ij} - \bar{X}_{ij}^{T} \Sigma_{ij}^{-1} \tau^2 \left(\Sigma_{ij} + \tau^2 I \right) \bar{X}_{ij}$$

$$= \bar{X}_{ij}^{T} \left[\left(\Sigma_{ij} + \tau^2 I \right)^{-1} + \Sigma_{ij}^{-1} \tau^2 \left(\Sigma_{ij} + \tau^2 I \right)^{-1} - \Sigma_{ij}^{-1} \tau^2 \left(\Sigma_{ij} + \tau^2 I \right)^{-1} \right]$$

(7.17)

$$\bar{X}_{ij} = \bar{X}_{ij}^{T} \left(\Sigma_{ij} + \tau^2 I \right)^{-1} \bar{X}_{ij}$$

and

$$\frac{1}{\tau^2} \eta_i^{T} \eta_i - \frac{1}{\tau^2} \eta_i^{T} \left(\Sigma_{ij}^{-1} + \tau^{-2}I \right)^{-1} \frac{1}{\tau^2} \eta_i$$

$$= \frac{1}{\tau^2} \eta_i^{T} \left(\Sigma_{ij} + \tau^2 I \right) \left(\Sigma_{ij} + \tau^2 I \right)^{-1} \eta_i - \frac{1}{\tau^2} \eta_i^{T} \Sigma_{ij} \left(\Sigma_{ij} + \tau^2 I \right)^{-1} \eta_i$$

(7.18)

$$= \eta_i^{T} \left(\Sigma_{ij} + \tau^2 I \right)^{-1} \eta_i$$

and further,

$$\bar{X}_{ij}^{T} \Sigma_{ij}^{-1} \left(\Sigma_{ij}^{-1} + \tau^{-2}I \right)^{-1} \frac{\eta_i}{\tau^2} = \bar{X}_{ij}^{T} \left(\Sigma_{ij} + \tau^2 I \right)^{-1} \eta_i$$

(7.19)

Finally, if we use (7.16)–(7.19) in (7.15), we get

$$P\left(\bar{X}_{ij}\right) = (2\pi)^{-k/2}\left|\Sigma_{ij} + \tau^2 I\right|^{-1}$$
$$\exp\left\{-\frac{1}{2}\left(\bar{X}_{ij} - \eta_i\right)^T \left(\Sigma_{ij} + \tau^2 I\right)^{-1} \left(\bar{X}_{ij} - \eta_i\right)\right\}$$

(7.20)

which shows that the unconditional distribution of $\bar{X}_{ij}$ is a multivariate normal with mean vector η_i and covariance matrix $\left(\Sigma_{ij} + \tau^2 I\right)$.

Although for simplicity, in the above derivation it was assumed that the litter means $\mu_{ij}(t_1),\ldots,\mu_{ij}(t_k)$ are independent, it is possible to relax that assumption. If we assume that the joint distribution of $\mu_{ij}(t_1),\ldots,\mu_{ij}(t_k)$ is a multivariate normal with mean vector $\eta_i = \left[\eta_i(t_1),\ldots,\eta_i(t_k)\right]^T$ and covariance matrix Ω, then with steps similar to above, we can prove that the unconditional distribution of $\bar{X}_{ij}$ is a multivariate normal with mean vector η_i and covariance matrix $\left(\Sigma_{ij} + \Omega\right)$. Note also that if it can be assumed that

$$\Sigma_{ij} = \Sigma_i;\ j = 1,\ldots,m_i,$$

that is, the covariance matrices are identical across litters, then analogous to (7.7) we have,

$$V\left(\bar{X}_{ij}\right) = \frac{1}{n_{ij}} V_{ij}\left(I + n_{ij}\theta_i\right)$$

where $\theta_i = \tau^2 \Sigma_i^{-1}$ is the overdispersion factor.

7.7 Generalization to Non-Symmetric Distributions

Up to this point, we have assumed that the distribution of the neurological variable from the DNT study is normal. As we have pointed out in this and in the last chapter, in general, normal or lognormal distributions appear to be adequate for continuous responses in toxicological experiments. However, as we have mentioned in the last chapter, it is not uncommon that the data from a DNT study would be non-normal. Also, as discussed earlier in this chapter, analysis of statistical problems arising from DNT studies are elaborated in detail by Holson et al. (2008). The methods of analysis of data mostly call for using techniques of factorial design and analysis of variance to detect dose-response effects based on the mean responses from different litters. Although these techniques are based on the assumptions of normality and homogeneity of variance, the paper mentions that as long as the distributions of dose groups have the same shape, i.e. same skewness and group sizes are reasonably similar, then due to robustness of the methodology, the analysis can be adopted except in extreme cases. The paper warns, however, that "treating multiple offspring from the same litter as independent is a fundamental violation of assumptions." It is well known that in developmental neurotoxicity experiments, responses from same litter tend to be more similar than responses from different litters. In this section, we study a modeling structure for the responses in a DNT study that could be non-symmetric and includes the symmetric normal distribution and a special case. Specifically, we describe the distribution of measurements within each litter as a skew-normal. This class of distributions has received an increasing amount of attention in recent years because of its generality and a wide range of applications. Expressing, once again, the location parameter of this distribution by a normal model, we develop a hierarchical structure for the unconditional distribution of the litter means. Much of the derivation is also described by Razzaghi (2014). First, we give a brief introduction to the skew-normal distribution.

7.7.1 Skew-Normal Distributions

A random variable X is said to have a skew-normal distribution with location parameter μ, scale parameter σ, and shape parameter λ when its density is given by

$$f\left(x \mid \mu, \sigma, \lambda\right) = \frac{2}{\sigma} \phi\left(\frac{x - \mu}{\sigma}\right) \Phi\left(\lambda \frac{x - \mu}{\sigma}\right) \tag{7.21}$$

where $\phi(\cdot)$ and $\Phi(\cdot)$, respectively, denote the density and the cumulative distribution functions of the standard normal. The skew-normal distribution was first introduced by Azzalini (1985). Since its introduction, the distribution has been studied by many researchers and a vast amount of literature describes the properties of this class of distributions. In a follow-up paper, Azzalini (2005) provides an informative overall account of the skew-normal models, discussing also the related multivariate families. A thorough account of the skew-normal distributions and other skew-elliptical distributions can be found in Genton (2004) and Arellano-Valle and Genton (2005). Note that the class of distributions (7.21) includes the normal as the special case of $\lambda = 0$ and produces the half-normal density when λ approaches infinity. Other interesting properties of the skew-normal distribution include the fact that the density is strongly unimodal; i.e. it is log-concave on its support (Sidak et al., 1999). Also, if $Z = (X - \mu)/\sigma$, then Z^2 has a χ^2 distribution with one degree of freedom and the moment generating function of Z is given by

$$M(t) = 2\exp(t^2/2)\Phi\left(\frac{\lambda t}{\sqrt{1 + \lambda^2}}\right) \tag{7.22}$$

Lin et al. (2007) showed that all moments of X may be derived recursively without using the moment generating function by applying a set of recursive equations. Specifically, the mean and variance of X are given by

$$E(X) = \mu + \sqrt{\frac{2}{\pi}} \frac{\lambda \sigma}{1 + \lambda^2} \qquad (7.23)$$

and

$$V(X) = \left(1 - \frac{2}{\pi} \frac{\lambda^2}{1 + \lambda^2} \right) \sigma^2 \qquad (7.24)$$

Henze (1986) showed that the even moments of Z coincide with those of the standard normal and derived an expression for all the odd moments. The problems of parameter estimation were discussed by Arnold et al. (1993) who derived moment estimators for μ, σ, and λ. Pewsey (2000) discussed the problems of inference for the univariate skew-normal distribution in length. He showed that because there are some difficulties in the estimation of parameters through the method of moments or the maximum likelihood method, direct parameterization should not be used as a general basis for estimation. The method of moments can lead to inadmissible estimates and the likelihood is not able to estimate a single value of the shape parameter among all possible values in small to moderate sample sizes. In fact, when the observations are all positive (or all negative), the likelihood is monotone increasing, and maximization of the likelihood produces ∞ as the estimate of the shape parameter λ. To resolve this problem, a reparameterization of the likelihood along with a nonlinear constrained iterative procedure is used. Another problem with the likelihood as described by Azzalini and Capitanio (1999) is that there is always an inflection point of the log-likelihood at $\lambda = 0$. This, in turn, means that the Fisher information matrix becomes singular when the shape parameter is zero. Sartori (2006) points out that the problems associated with the maximum likelihood estimation is confined only to the shape parameter and empirical evidence shows that maximum likelihood estimates of the location and scale parameters always exist. The Bayesian inference for the skewness parameter of the skew-normal distribution was discussed by Bayes and Branco (2007) and Arellano-Valle et al. (2009). The Bayesian inference was extended to the general class of skew-symmetric distributions by Arellano-Valle et al. (2008) and Branco et al. (2013).

7.7.2 Unconditional Distribution of Litter Means

Let $X_{ij1}, X_{ij2}, \ldots, X_{ijn_{ij}}$ be the measurements on the n_{ij} pups in the jth litter of the ith dose group. We remind the reader that in most cases n_{ij} is 2 or 4 and that each sex is analyzed separately. Suppose that the distribution of $X_{ijl}; l = 1, \ldots, n_{ij}$ be skew-normal with mean μ_{ij}, variance σ_{ij}^2, and shape parameter λ_{ij}. Let $\bar{X}_{ij}$ be the mean of measurements within the jth litter of the ith dose group. Thus,

$$\bar{X}_{ij} = \sum_{l=1}^{n_{ij}} X_{ijl}.$$

Then, from the result of the study by Chen et al. (2004) we have that the density of the conditional distribution of $\bar{X}_{ij}$ is given by

$$f(\bar{x}_{ij} \mid \mu) = 2^{n_{ij}} \sqrt{\frac{n_{ij}}{2\pi}} \exp\left\{-\frac{n_{ij}}{2}\left(\frac{\bar{x}_{ij} - \mu_{ij}}{\sigma_{ij}}\right)^2\right\} \Phi_{n_{ij}}(W_{ij}),$$

where

$$W_{ij} = \lambda_{ij}\left(\frac{\bar{x}_{ij} - \mu_{ij}}{\sigma_{ij}}\right)\left[\frac{1}{1+\lambda_{ij}^2} I_{n_{ij}} + \frac{1}{n_{ij}\left(1+\lambda_{ij}^2\right)} 1_{n_{ij}} 1'_{n_{ij}}\right]^{\frac{1}{2}} 1_{n_{ij}}.$$

Here, $1_{n_{ij}} = (1,1,\ldots,1)'$ is the n_{ij}-vector of 1's, $I_{n_{ij}}$ is the n_{ij}-dimensional identity matrix and $\Phi_{n_{ij}}(.)$ is the n-variate cdf of the standard normal random vector. Now, as before, we assume that μ_{ij} has a normal distribution with mean η_i and standard deviation τ_i. Then, using the result of the study by Liseo and Loperfido (2006) we can prove that the unconditional distribution of $\bar{X}_{ij}$ is given by

$$f(\bar{x}_{ij}) = \left[2\pi\left(\frac{\sigma_{ij}^2}{n_{ij}} + \tau_{ij}^2\right)\right]^{-\frac{1}{2}} \exp\left\{-\frac{1}{2\left(\frac{\sigma_{ij}^2}{n_{ij}} + \tau_{ij}^2\right)}(\bar{x}_{ij} - \eta_i)^2\right\}$$

$$\Phi_{n_{ij}}\left\{\lambda_{ij}\left(I_{n_{ij}} + \frac{\lambda_{ij}^2}{n_{ij} + \frac{\sigma_{ij}^2}{\tau_{ij}^2}} A_{n_{ij}} A'_{n_{ij}}\right)^{-\frac{1}{2}}\left[\frac{\bar{x}_{ij} - u}{\sigma_{ij}} - \frac{1}{\sqrt{n_{ij} + \frac{\sigma_{ij}^2}{\tau_{ij}^2}}}\right] A_{n_{ij}}\right\}$$

(7.25)

where

$$A_{n_{ij}} = \left(\frac{1}{1+\lambda_{ij}^2} I_{n_{ij}} + \frac{1}{n_{ij}\left(1+\lambda_{ij}^2\right)} 1_{n_{ij}} 1'_{n_{ij}}\right)^{\frac{1}{2}} 1_{n_{ij}} \qquad (7.26)$$

and u is the realized value of the random variable U_{ij} defined as

$$U_{ij} = Z + \frac{\lambda_{ij}}{\sqrt{n_{ij} + \dfrac{\sigma_{ij}^2}{\tau_{ij}^2}}} A_{n_{ij}} Y \qquad (7.27)$$

with $\mathbf{Z} = \left(Z_1, Z_2, \ldots, Z_{n_{ij}}\right)^T$ being a vector of n_{ij}-dimensional independent standard normal variables and Y being a standard normal variable. For details of the proof we refer to Razzaghi (2014). It is possible to show (see Exercise 4 at the end of this chapter) that as the shape parameter λ_{ij} approaches 0, the distribution of $\bar{X}_{ij}$ becomes normal given by (7.4). This shows that (7.25) is a generalization of (7.4) and allows non-symmetric distributions for the response variable and includes normality as a special case.

Now, from (7.23) and (7.24), we have

$$E\left(\bar{X}_{ij}\right) = E\left[E\left(\bar{X}_{ij} \mid \mu_{ij}\right)\right] = \eta_i + \sqrt{\frac{2}{\pi}} \frac{\lambda_{ij}\sigma_{ij}}{\sqrt{1 + \lambda_{ij}^2}}$$

and

$$V\left(\bar{X}_{ij}\right) = E\left[V\left(\bar{X}_{ij} \mid \mu_{ij}\right)\right] + V\left[E\left(\bar{X}_{ij} \mid \mu_{ij}\right)\right]$$

$$= E\left[\left(1 - \frac{2}{\pi}\frac{\lambda_{ij}^2}{1 + \lambda_{ij}^2}\right)\frac{\sigma_{ij}^2}{n_{ij}}\right] + V\left[\mu_{ij} + \sqrt{\frac{2}{\pi}}\frac{\lambda_{ij}\sigma_{ij}}{\sqrt{1 + \lambda_{ij}^2}}\right] \qquad (7.28)$$

$$= \frac{\sigma_{ij}^2}{n_{ij}}\left(1 - \frac{2}{\pi}\frac{\lambda_{ij}^2}{1 + \lambda_{ij}^2}\right) + \tau_{ij}^2$$

Further, the moment generating function of $\bar{X}_{ij}$ is given by

$$M_{\bar{X}_{ij}}(t) = E\left[e^{t\bar{X}_{ij}}\right] = \prod_{l=1}^{n_{ij}} E\left[e^{\frac{t}{n_{ij}}X_{ijl}}\right].$$

From (7.22),

$$E\left[e^{t\bar{X}_{ij}} \mid \mu_{ij}\right] = e^{\mu_{ij}} M(\sigma_{ij}t)$$

and since

$$E\left(e^{\mu_{ij}}\right) = e^{\eta_i t + \frac{\tau_{ij}^2 t^2}{2}}$$

we have

$$E\left[e^{tX_{ijl}}\right] = E\left[e^{\mu_{ij}} M(\sigma_{ij}t)\right] = E\left(e^{\mu_{ij}}\right) M(\sigma_{ij}t)$$

$$= 2\exp\left\{\eta_i t + \frac{\tau_{ij}^2 t^2}{2} + \frac{\sigma_{ij}^2 t^2}{2}\right\} \Phi\left(\frac{\lambda_{ij}\sigma_{ij}}{\sqrt{1+\lambda_{ij}^2}} t\right)$$

from which we get

$$M_{\bar{X}_{ij}}(t) = 2^n \prod_{i=1}^{n} \exp\left(\frac{\eta_i t}{n_{ij}} + \frac{\tau_{ij}^2 t^2}{2n_{ij}^2} + \frac{\sigma_{ij}^2 t^2}{2n^2}\right) \Phi\left(\frac{\sigma_{ij}}{n_{ij}} \frac{\lambda_{ij}}{\sqrt{1+\lambda_{ij}^2}} t\right)$$

$$= 2^n \exp\left\{\eta_i t + \left(\sigma_{ij}^2 + \tau_{ij}^2\right) \frac{t^2}{2n_{ij}}\right\} \left[\Phi\left(\frac{\sigma_{ij}}{n_{ij}} \frac{\lambda_{ij}}{\sqrt{1+\lambda_{ij}^2}} t\right)\right]^{n_{ij}}.$$

7.7.3 Intralitter Correlation

From (7.5) and (7.28) we have

$$\frac{\sigma_{ij}^2}{n_{ij}}\left(1 - \frac{2}{\pi}\frac{\lambda_{ij}^2}{1+\lambda_{ij}^2}\right) + \tau_{ij}^2 = \frac{\sigma_{ij}^2}{n_{ij}}\left(1 - \frac{2}{\pi}\frac{\lambda_{ij}^2}{1+\lambda_{ij}^2}\right) + \frac{\tau_{ij}^2}{n_{ij}} + \frac{n_{ij}\left(n_{ij}-1\right)}{n_{ij}^2} \mathrm{Cov}\left(X_{ijl}, X_{ijl'}\right) \ l \neq l'$$

which gives

$$\mathrm{Cov}\left(X_{ijl}, X_{ijl'}\right) = \tau_{ij}^2.$$

Hence, the intralitter correlation for the *j*th litter of the *i*th dose group can be expressed as

$$\rho_{ij} = \frac{\tau_{ij}^2}{\sigma_{ij}^2 + \tau_{ij}^2 - \sigma_{ij}^2 \frac{2}{\pi}\frac{\lambda_{ij}^2}{1+\lambda_{ij}^2}} = \frac{\tau_{ij}^2}{\sigma_{ij}^2\left(1 - \frac{2}{\pi}\frac{\lambda_{ij}^2}{1+\lambda_{ij}^2}\right) + \tau_{ij}^2}.$$

If we let $\theta_{ij} = \frac{\rho_{ij}}{1-\rho_{ij}}$, then $\theta_{ij} = \frac{\tau_{ij}^2}{\sigma_{ij}^2\left(1 - \frac{2}{\pi}\frac{\lambda_{ij}^2}{1+\lambda_{ij}^2}\right)}$, and substituting in (7.17), we get

$$V\left(\bar{X}_{ij}\right) = \frac{\sigma_{ij}^2}{n_{ij}}\left(1 - \frac{2}{\pi}\frac{\lambda_{ij}^2}{1+\lambda_{ij}^2}\right)\left(1 + n_{ij}\theta_{ij}\right) \qquad (7.29)$$

which shows that θ_{ij} can be interpreted as the overdispersion parameter. Note that as the shape parameter λ_{ij} approaches 0, that is when $\lambda_{ij} \to 0$, (7.29) becomes equivalent to (7.7).

Example 7.4

We return once again to the representative data set generated by Razzaghi and Kodell (2004) from the experiment to assess the neurodevelopmental effects of a mixture of 2,4-D and 2,4,5-T described in Mohammad and St. Omer (1985, 1986). The data set was used for demonstrating the risk assessment methodology based on the normal-normal structure developed in Section 7.5 of this chapter. However, using the Shapiro-Wilk test of normality, Razzaghi (2014) shows that the normality assumption for the data is unreasonable. Assuming the skew-normal distribution with equal shape parameter λ and common variance σ^2 across all litters for the responses, Razzaghi (2014) finds that that the moment estimates of λ and σ were, respectively, 0.514 and 32.115 for the female pups and 0.980 and 42.345 for the male pups. Calculating $V\left(\bar{X}_{ij}\right)$ from (7.17), this leads to reductions of 68.71 and 79.62 in the variance of female and male pups, respectively.

Exercises

1. In a DNT study, it has been determined that the pup neurological data do not exhibit symmetry and thus a normal distribution is not appropriate. Suppose that the distribution of responses within jth litter of the ith dose group follow an exponential distribution with parameter λ_{ij}

$$f\left(x_{ijl} \mid \lambda_{ij}\right) = \lambda_{ij} e^{-\lambda_{ij} x_{ijl}} \quad x_{ijl} \geq 0$$

   a. Show that the distribution of the sample mean $\bar{X}_{ij}$ within the jth litter of the ith dose group is a gamma distribution with shape parameter n_{ij} and rate parameter $\dfrac{\lambda_{ij}}{n_{ij}}$.

   b. Assuming that litter effect can be expressed as a gamma distribution on λ_{ij} with parameters α_{0i} and β_{0i} for the ith dose group, derive the unconditional distribution of $\bar{X}_{ij}$ and show that it belongs to the gamma family.

2. York et al. (2004) describe an experiment for assessing the adverse neurodevelopmental effects ammonium perchlorate in drinking water in Sprague-Dawley rats. Pregnant female dams were

exposed to saline and four non-zero doses of 0.1, 1.0, 3.0, and 10.0 mg/kg-day of ammonium perchlorate in drinking water with 25 dams per group. Exposure occurred from gestation day 0 to lactation day 10. On lactation day 10, the dose group sizes were reduced from 25 dams to 20 mated females by first sacrificing the nonpregnant rats and then culling rats with the lowest number of litters. This was done for the purpose of making sure that the litter sizes were sufficiently large. Litters were culled to 8 pups with 4 males and 4 females randomly selected from each litter. There were therefore 400 male and 400 female pups divided into four subsets for a variety of testing and measurements. Table 7.8 below gives the passive avoidance results for latency trial on postpartum day 23 (session 1) and one week later (session 2) for both male and female pups.

Using the data from Session I for both male and female pups, mimic the calculations in Examples 7.1 and 7.2. Thus perform the following:

a. Generate the data for the observed values with one pup per sex per litter per dose group. Note that in Tables 7.8, standard deviations are given rather than SEM's as in Table 7.2.

b. Test for dose effect using analysis of variance

c. Assuming a hierarchical normal structure, calculate maximum likelihood estimates of the model parameters

d. Calculate the BMD's and BMDL's for both male and female pups for a given risk of $\pi_0 = 0.05$.

3. Consider now Passive Avoidance measurements data from last exercise from both Session I and Session II for both male and female pups. Use the hierarchical multivariate normal structure described in Section 6 of this chapter. Since there appears to be a high degree of variability among the means, suppose that litter means $\mu_{ij}(t_1), \mu_{ij}(t_2)$ $j = 1, \ldots, 20$, $i = 0, \ldots, 4$ are independent, but have unequal variances. Therefore let the distribution of $\mu_{ij} = \left[\mu_{ij}(t_1), \mu_{ij}(t_2) \right]^T$ be a multivariate normal with mean $\eta_i = \left[\eta_i(t_1), \eta_i(t_2) \right]^T$ and the covariance matrix $\Omega = \text{diag}\left(\tau_1^2, \tau_2^2 \right)$. Find the unconditional distribution of $\bar{X}_{ij}$ and find the probability that for the male pups the sum of the litter means for the two sessions exceeds 25.

4. Using equations (7.15) and (7.16), show that as $\lambda_{ij} \rightarrow 0$, then (7.14) approaches (7.4) i.e. when the shape parameter is zero, then the distribution of $\bar{X}_{ij}$ is normal. Thus (7.14) is a generalization of (7.4) to include a more flexible class of distributions and includes normality as a special case.

TABLE 7.8

Passive Avoidance in Rat Pups

Dose (mg/kg-day)	0		0.1		1		3		10	
	Mean	SD	Mean	SD	Mean	SD	Mean	SD	Mean	SD
Male										
Session I	7.2	5	6.8	3.6	7	4.6	7.1	3.4	5.9	2.5
Session II	22.2	21.2	27.2	20.8	30.4	24.2	27.2	16.6	24.6	18.2
Female										
Session I	6.8	4.6	8.3	4.8	7.2	3.6	8	3.5	8	6.6
Session II	22.9	17.6	24.3	19.7	28.4	21.9	25.4	24.3	20.8	17.1

Source: Data from York et al. (2004).

References

Arellano-Valle, R. B. and Genton, M. G. (2005). On fundamental skew distributions. *Journal of Multivariate Analysis* 96, 93–116.

Arellano-Valle, R. B., Castro, L. M., Genton, M. G., and Gomez, H. W. (2008). Bayesian inference for shape mixture of skewed distributions, with application to regression analysis. *Bayesian Analysis* 3, 513–540.

Arellano-Valle, R. B., Genton, M. G., and Loschi, R. H. (2009). Shape mixtures of multivariate skew-normal distributions. *Journal of Multivariate Analysis* 100, 91–101.

Arnold, B. C., Beaver, R. J., Groenveld, R. A., and Meeker, W. Q. (1993). The non-truncated marginal of a truncated bivariate normal distribution. *Psychometrica* 58, 471–478.

Azzalini, A. (1985). A class of distribution which includes the normal ones. *Statistica* 46, 199–208.

Azzalini, A. (2005). The skew-normal distribution and related multivariate families (with discussion). *Scandinavian Journal of Statistics* 32, 159–188.

Azzalini, A. and Capitanio, A. (1999). Statistical applications of the multivariate skew normal distributions. *Journal of Royal Statistical Society B* 61, 579–602.

Bal-Price, A. and Fritsche, E. (2018). Developmental neurotoxicity (Editorial). *Toxicology and Applied Pharmacology* 354, 1–2.

Barnes, D. B. and Dourson, M. (1989). Reference dose (RfD): Description and use in human health risk assessment. *Regulatory Toxicology and Pharmacology* 8, 471–486.

Bayes, C. L. and Branco, M. D. (2007). Bayesian inference for the skewness parameter of the scalar skew-normal distribution. *Brazilian Journal of Probability and Statistics* 21, 141–163.

Behl, M., Ryan, K., Hsieh, J.-H., Parham, F., Shapiro, A. J., Collins, B. J., Sipes, N. S., Birnbaum, L. S., Bucher, J. R., and Foster, P. M. D. (2019). Screening for developmental neurotoxicity at the National Toxicology Program: The future is here. *Toxicological Sciences* 167, 6–14.

Branco, M. D., Genton, M. G., and Liseo, B. (2013). Objective Bayesian analysis of skew-t distributions. *Scandinavian Journal of Statistics* 40, 63–85.

Catalano, P. J. and Ryan, L. M. (1992). Bivariate latent variable models for clustered discrete and continuous outcomes. *Journal of the American Statistical Association* 87, 651–658.

Chang, L. W. and Guo, G. L. (1998). Fetal minamata disease—Congenital methylmercury poisoning. In *Handbook of developmental neurotoxicology*, W. L. Slikker and L. W. Chang eds. Academic Press, London.

Chen, J. T., Gufta, A. K., and Nguyen, T. T. (2004). The density of the skew normal sample mean and its application. *Journal of Statistical Simulation* 74, 484–494.

Chen, J. J., Kodell, R. L., and Gaylor, D. W. (1996). Risk assessment for non-quantal toxic effects. In *Toxicology and risk assessment: Principles, methods and applications*, A. M. Fan and L. W. Chang eds. Marcel Dekker, New York, pp. 503–513. Academic Press, New York, pp. 507–515.

Cohen, J. (1988). *Statistical power analysis for the behavioral sciences*, 2nd edition. Lawrence Earlbaum Associates, Hillsdale.

DuboVický, M., Kovacovský, P., Ujházy, E., Navarová, J., Brucknerová, I., and Mach, M. (2008). Evaluation of developmental neurotoxicity: Some important issues focused on neurobehavioral development. *Interdisciplinary Toxicology* 1, 206–210.

Gaylor, D. W. (1994). Dose-response modeling. In *Developmental toxicology*, 2nd edition. C. A. Kimmel and J. Buelke-Sam eds. Raven Press, New York, pp. 363–375.

Gaylor, D. W. and Slikker, W. L. (1990). Risk assessment for neurotoxic effects. *Neurotoxicology* 11, 211–218.

Giordano, G. and Costa, L. G. (2012). Developmental neurotoxicity: Some old and new issues. International Scholarly Research Network, *ISRN Toxicology*, 2012, Article ID 814795.

Genton, M. (2004). *Skew-elliptical distributions and their applications: A journey beyond normality*. Chapman & Hall/CRC, Boca Raton, FL.

Gosden, C. (1998). We've got to try to do something. USAWeekend, May 15–17, pp. 14–16.

Henze, N. (1986). A probabilistic representation of the "skew-normal" distribution. *Scandinavian Journal of Statistics* 13, 271–275.

Holson, R. R., Freshwater, L., Maurissen, J. P. J., Moser, V. C., and Phang, W. (2008). Statistical issues and techniques appropriate for developmental neurotoxicity testing. *Neurotoxicology and Teratology* 30, 326–348.

Kavlock, R. J., Allen, B. C., Faustman, E. M., and Kimmel, C. A. (1995).Dose-response assessments for developmental toxicity. *Fundamental and Applied Toxicology* 26, 211–222.

Kodell, R. L., Howe, R. B., Chen, J. J., and Gaylor, D. W. (1991). Mathematical modeling of reproductive and developmental toxic effects for quantitative risk assessment. *Risk Analysis* 11, 583–590.

Kodell, R. L. and West, R. W. (1993). Upper confidence limits on excess risk for quantitative responses. *Risk Analysis* 13, 177–182.

Lin, T. I., Lee, J. C., and Yen, S. Y. (2007). Finite mixture modeling using the skew normal distribution. *Statistica Sinica* 17, 909–927.

Liseo, B. and Loperfido, N. (2006). A note on reference priors for the scalar skew-normal distribution. *Journal of Statistical Planning and Inference* 136, 373–389.

Mohammad, F. K. and St. Omer, V. (1985). Developing rat brain monoamine levels following in utero exposure to mixture of 2,4-dichlorophenoxyacetic and 2,4,5-trichlorophenoxyacetic acids. *Toxicology Letters* 29, 215–233.

Mohammad, F. K. and St. Omer, V. (1986). Behavioral and developmental effects in rats following in utero exposure to 2,4D/2,4,5-T mixture. *Nuerobehavioral Toxicology and Teratology* 8, 551–560.

Organization for Economic Cooperation and Development. (2003). OECD Environment, Health and Safety Publications Series on Testing and Assessment No. 20. Guidance Document for Neurotoxicity Testing. Environment Directorate, OECD, Paris.

Organization for Economic Cooperation and Development. (2006). OECD Guideline for the Testing of Chemicals. Draft Test Guideline 426: Developmental Neurotoxicity Study.

Organization for Economic Cooperation and Development. (2011). Extended one-generation reproductive toxicity study. *OECD Guideline for Testing of Chemicals*, No. 443, OECD, Paris.

Pewsey, A. (2000). Problems of inference for Azzalini's skew-normal distribution. *Journal of Applied Statistics* 27, 859–870.

Prentice, R. L. (1986). Binary regression using an extended beta-binomial distribution, with discussion of correlation induced by covariate measurement errors. *Journal of American Statistical Association* 81, 321–327.

Razzaghi, M. and Kodell, R. L. (2000). Dose-response modeling for developmental neurotoxicity data. *Environmental and Ecological Statistics* 7, 191–203.

Razzaghi, M. and Kodell, R. L. (2004). Quantitative risk assessment for developmental neurotoxic effects. *Risk Analysis* 24, 1673–1681.

Razzaghi, M. (2014). A hierarchical model for the skew-normal distribution with application in developmental neurotoxicology. *Communications in Statistics – Theory and Methods* 43, 1859–1872.

Rees, D. C., Francis, E. Z., and Kimmel, C. A. (1990). Scientific and regulatory issues relevant to assessing risk for developmental neurotoxicity: An overview. *Neurotoxicology and Teratology* 12, 175–181.

Ryan, L. M., Catalano, P. J., Kimmel, C. A., and Kimmel, G. L. (1991). Relationship between fetal weight and malformation in developmental toxicity studies. *Teratology* 44, 215–223.

Sartori, N. (2006). Bias prevention of maximum likelihood estimates for scalar skew normal and skew t distribution. *Journal of Statistical Planning and Inference* 136, 4259–4275.

Searle, S. R. (1971). *Linear Models*, Wiley, New York.

Sidak, Z., Hajek, J., and Sen, P. K. (1999). *Theory of rank tests*, 2nd edition. Academic Press, New York.

Slikker, W. and Chang, L. W. (1998). *Handbook of developmental neurotoxicology*. Academic Press, London.

Smirnova, L., Hogberg, H. T., Leist, M., and Hartung, T. (2014). Developmental neurotoxicity – challenges in the 21st century and in vitro opportunities. *Alternatives and Animal Use in the Life Sciences (ALTEX)* 31, 129–156.

Snedecor, G. W. and Cochran, W. G. (1980). *Statistical methods*, 7th edition. State University Press, Ames.

Stanton, M. E. and Spear, L. P. (1990). Workshop on the qualitative and quantitative comparability of human and animal developmental neurotoxicity. Work Group I report: Comparability of measures of developmental neurotoxicity in humans and laboratory animals. *Neurotoxicology and Teratology* 12, 261–267.

Tilson, H. A. (2000). Neurotoxicology risk assessment guidelines. *Neurotoxicology* 21, 189–194.

U.S. Environmental Protection Agency. (1998). Health Effects Test Guidelines, OPPTS 8706300. Developmental Neurotoxicity Study, EPA 712-C-98-239.

U.S. Environmental Protection Agency. (2016). North American Free Trade Agreement (NAFTA) Technical Working Group on Pesticides (TWG). Developmental Neurotoxicity Study Guidance Document.

Werboff, J. and Gottlieb, J. S. (1963). Drugs in pregnancy: Behavioral teratology. *Obstetrical and Gynecological Survey* 18, 420–423.

York, G. Y., Barnett Jr., J., Brown, W. R., Garman, R. H., Mattie, D. R., and Dodd, D. (2004). A rat neurodevelopmental evaluation of offspring, including evaluation of adult and neonatal thyroid, from mothers treated with ammonium perchlorate in drinking water. *International Journal of Toxicology* 23, 191–214.

Appendix 1: Data Generation for Mixture of Five Chemicals (Example 3.3)

In Chapter 3, we presented a methodology due to Gennings (1995) to assess the additivity of the effect of a mixture of chemicals. To illustrate the application of the methodology, in Example 3.3, we examined the additivity of the effect of a mixture of five polycyclic aromatic hydrocarbons. The example referred to a study described in Nesnow et al. (1998) and conducted at the Medical College of Ohio and Ohio State University. The five chemicals under study were benzo[a]pyrene (B(a)P), benzo[b]fluoranthene (B(b)F), dibenz[a,h] anthracene (DB(a,h)A), 5-methylchrysene (5MCHR), and cyclopenta[c,d] pyrene (CPP). The major outcome of interest in the study was the number of tumors on the lung of the surviving mice after eight months on strain A/J of the surviving mice. Since a summary data giving the mean and variance of the number of tumors along with the sample size were provided by Gennings (1995) and Nesnow et al. (1998), for the purpose of illustration it was necessary to generate a representative sample with a site-specific data set that provides the number of tumors for each rat. Here, we describe the methodology for generating that representative data set.

Denote by $\bar{X}_0$ and S_0^2 the mean and variance of the control group, and let $\bar{X}_{ij}$ and S_{ij}^2 be respectively the mean and variance of the jth dose level of the ith chemical for $j = 0, \ldots, m_i, 0 = 1, \ldots, 5$ whereby convention $\bar{X}_{i0} = \bar{X}_{0j} = \bar{X}_0$ and $S_{i0}^2 = S_{0j}^2 = S_0^2$. From Table 3.4, note that $m_1 = 6$, $m_2 = 5$, $m_3 = 4$, $m_4 = 4$, $m_5 = 4$. The boundaries

$$\alpha_{ij} = \bar{X}_{ij} - \frac{S_{ij}}{\sqrt{12}}$$

and

$$\beta_{ij} = \bar{X}_{ij} + \frac{S_{ij}}{\sqrt{12}}$$

were first calculated, and a discrete uniform random variable was then generated in the range $(\alpha_{ij}, \beta_{ij})$ in Excel. The process was repeated for the given sample size. Table A.1 displays the simulated data set.

TABLE A.1

A Simulated Representative Sample for the Number of Tumors in Mice Exposed to Polycyclic Aromatic Hydrocarbons

Control	Mean	Variance	Sample	Response	Dose				
					B(a)P	B(b)F	DB(a,h)A	5MCHR	CPP
Control	0.6	0.358	1	1	0	0	0	0	0
			2	1	0	0	0	0	0
			3	1	0	0	0	0	0
			4	1	0	0	0	0	0
			5	1	0	0	0	0	0
			6	1	0	0	0	0	0
			7	1	0	0	0	0	0
			8	1	0	0	0	0	0
			9	1	0	0	0	0	0
			10	0	0	0	0	0	0
			11	1	0	0	0	0	0
			12	1	0	0	0	0	0
			13	0	0	0	0	0	0
			14	1	0	0	0	0	0
			15	1	0	0	0	0	0
			16	1	0	0	0	0	0
			17	1	0	0	0	0	0
			18	0	0	0	0	0	0
			19	1	0	0	0	0	0
			20	0	0	0	0	0	0
B(a)P	0.45	0.682	1	1	5	0	0	0	0
			2	1	5	0	0	0	0

(*Continued*)

TABLE A.1 (CONTINUED)

A Simulated Representative Sample for the Number of Tumors in Mice Exposed to Polycyclic Aromatic Hydrocarbons

Control	Mean	Variance	Sample	Response	Dose				
					B(a)P	B(b)F	DB(a,h)A	5MCHR	CPP
			3	1	5	0	0	0	0
			4	0	5	0	0	0	0
			5	0	5	0	0	0	0
			6	0	5	0	0	0	0
			7	1	5	0	0	0	0
			8	0	5	0	0	0	0
			9	0	5	0	0	0	0
			10	0	5	0	0	0	0
			11	0	5	0	0	0	0
			12	1	5	0	0	0	0
			13	1	5	0	0	0	0
			14	0	5	0	0	0	0
			15	0	5	0	0	0	0
			16	1	5	0	0	0	0
			17	0	5	0	0	0	0
			18	0	5	0	0	0	0
			19	1	5	0	0	0	0
			20	1	5	0	0	0	0
	0.53	0.64	1	1	10	0	0	0	0
			2	0	10	0	0	0	0
			3	0	10	0	0	0	0
			4	1	10	0	0	0	0

(*Continued*)

TABLE A.1 (CONTINUED)

A Simulated Representative Sample for the Number of Tumors in Mice Exposed to Polycyclic Aromatic Hydrocarbons

Control	Mean	Variance	Sample	Response	Dose				
					B(a)P	B(b)F	DB(a,h)A	5MCHR	CPP
			5	1	10	0	0	0	0
			6	1	10	0	0	0	0
			7	1	10	0	0	0	0
			8	1	10	0	0	0	0
			9	1	10	0	0	0	0
			10	1	10	0	0	0	0
			11	1	10	0	0	0	0
			12	0	10	0	0	0	0
			13	1	10	0	0	0	0
			14	0	10	0	0	0	0
			15	0	10	0	0	0	0
			16	1	10	0	0	0	0
			17	0	10	0	0	0	0
	4.37	7.91	1	4	50	0	0	0	0
			2	5	50	0	0	0	0
			3	5	50	0	0	0	0
			4	4	50	0	0	0	0
			5	5	50	0	0	0	0
			6	4	50	0	0	0	0
			7	4	50	0	0	0	0
			8	4	50	0	0	0	0
			9	4	50	0	0	0	0

(Continued)

TABLE A.1 (CONTINUED)

A Simulated Representative Sample for the Number of Tumors in Mice Exposed to Polycyclic Aromatic Hydrocarbons

Control	Mean	Variance	Sample	Response	B(a)P	B(b)F	DB(a,h)A	5MCHR	CPP
			10	4	50	0	0	0	0
			11	5	50	0	0	0	0
			12	4	50	0	0	0	0
			13	5	50	0	0	0	0
			14	4	50	0	0	0	0
			15	5	50	0	0	0	0
			16	4	50	0	0	0	0
			17	4	50	0	0	0	0
			18	5	50	0	0	0	0
			19	5	50	0	0	0	0
	7.14	14.8	1	7	75	0	0	0	0
			2	8	75	0	0	0	0
			3	7	75	0	0	0	0
			4	8	75	0	0	0	0
			5	8	75	0	0	0	0
			6	7	75	0	0	0	0
			7	7	75	0	0	0	0
	12.7	19.5	1	13	100	0	0	0	0
			2	12	100	0	0	0	0
			3	14	100	0	0	0	0
			4	13	100	0	0	0	0
			5	12	100	0	0	0	0

(Continued)

TABLE A.1 (CONTINUED)

A Simulated Representative Sample for the Number of Tumors in Mice Exposed to Polycyclic Aromatic Hydrocarbons

Control	Mean	Variance	Sample	Response	Dose				
					B(a)P	B(b)F	DB(a,h)A	5MCHR	CPP
			6	13	100	0	0	0	0
			7	13	100	0	0	0	0
			8	12	100	0	0	0	0
			9	12	100	0	0	0	0
			10	14	100	0	0	0	0
			11	14	100	0	0	0	0
			12	12	100	0	0	0	0
			13	13	100	0	0	0	0
			14	12	100	0	0	0	0
			15	13	100	0	0	0	0
			16	12	100	0	0	0	0
	33	109.2	1	33	200	0	0	0	0
			2	35	200	0	0	0	0
			3	34	200	0	0	0	0
			4	31	200	0	0	0	0
			5	33	200	0	0	0	0
			6	33	200	0	0	0	0
			7	34	200	0	0	0	0
			8	34	200	0	0	0	0
			9	32	200	0	0	0	0
			10	33	200	0	0	0	0
			11	31	200	0	0	0	0

(Continued)

TABLE A.1 (CONTINUED)

A Simulated Representative Sample for the Number of Tumors in Mice Exposed to Polycyclic Aromatic Hydrocarbons

Control	Mean	Variance	Sample	Response	Dose				
					B(a)P	B(b)F	DB(a,h)A	5MCHR	CPP
			12	35	200	0	0	0	0
			13	36	200	0	0	0	0
			14	32	200	0	0	0	0
			15	31	200	0	0	0	0
			16	31	200	0	0	0	0
			17	33	200	0	0	0	0
			18	33	200	0	0	0	0
			19	33	200	0	0	0	0
			20	30	200	0	0	0	0
			21	33	200	0	0	0	0
			22	32	200	0	0	0	0
			23	35	200	0	0	0	0
			24	34	200	0	0	0	0
B(b)F	0.67	0.588	1	1	0	10	0	0	0
			2	1	0	10	0	0	0
			3	0	0	10	0	0	0
			4	1	0	10	0	0	0
			5	1	0	10	0	0	0
			6	1	0	10	0	0	0
			7	1	0	10	0	0	0
			8	1	0	10	0	0	0
			9	1	0	10	0	0	0

(Continued)

TABLE A.1 (CONTINUED)

A Simulated Representative Sample for the Number of Tumors in Mice Exposed to Polycyclic Aromatic Hydrocarbons

Control	Mean	Variance	Sample	Response	Dose				
					B(a)P	B(b)F	DB(a,h)A	5MCHR	CPP
			10	1	0	10	0	0	0
			11	1	0	10	0	0	0
			12	1	0	10	0	0	0
			13	1	0	10	0	0	0
			14	1	0	10	0	0	0
			15	1	0	10	0	0	0
			16	1	0	10	0	0	0
			17	1	0	10	0	0	0
			18	1	0	10	0	0	0
	2	3.47	1	3	0	50	0	0	0
			2	2	0	50	0	0	0
			3	2	0	50	0	0	0
			4	2	0	50	0	0	0
			5	2	0	50	0	0	0
			6	2	0	50	0	0	0
			7	2	0	50	0	0	0
			8	2	0	50	0	0	0
			9	2	0	50	0	0	0
			10	2	0	50	0	0	0
			11	2	0	50	0	0	0
			12	2	0	50	0	0	0
			13	1	0	50	0	0	0

(Continued)

TABLE A.1 (CONTINUED)

A Simulated Representative Sample for the Number of Tumors in Mice Exposed to Polycyclic Aromatic Hydrocarbons

Control	Mean	Variance	Sample	Response	Dose				
					B(a)P	B(b)F	DB(a,h)A	5MCHR	CPP
			14	2	0	50	0	0	0
			15	2	0	50	0	0	0
			16	2	0	50	0	0	0
			17	2	0	50	0	0	0
			18	2	0	50	0	0	0
			19	2	0	50	0	0	0
			20	2	0	50	0	0	0
	4.1	1.43	1	4	0	75	0	0	0
			2	4	0	75	0	0	0
			3	4	0	75	0	0	0
			4	4	0	75	0	0	0
			5	4	0	75	0	0	0
			6	4	0	75	0	0	0
			7	4	0	75	0	0	0
			8	4	0	75	0	0	0
			9	4	0	75	0	0	0
			10	4	0	75	0	0	0
	5.3	10.8	1	5	0	100	0	0	0
			2	5	0	100	0	0	0
			3	5	0	100	0	0	0
			4	5	0	100	0	0	0
			5	4	0	100	0	0	0

(Continued)

TABLE A.1 (CONTINUED)

A Simulated Representative Sample for the Number of Tumors in Mice Exposed to Polycyclic Aromatic Hydrocarbons

Control	Mean	Variance	Sample	Response	Dose				
					B(a)P	B(b)F	DB(a,h)A	5MCHR	CPP
			6	5	0	100	0	0	0
			7	6	0	100	0	0	0
			8	6	0	100	0	0	0
			9	5	0	100	0	0	0
			10	6	0	100	0	0	0
			11	5	0	100	0	0	0
			12	5	0	100	0	0	0
			13	5	0	100	0	0	0
			14	5	0	100	0	0	0
			15	6	0	100	0	0	0
			16	5	0	100	0	0	0
			17	4	0	100	0	0	0
			18	5	0	100	0	0	0
			19	5	0	100	0	0	0
			20	6	0	100	0	0	0
	6.95	13.1	1	6	0	200	0	0	0
			2	6	0	200	0	0	0
			3	8	0	200	0	0	0
			4	7	0	200	0	0	0
			5	6	0	200	0	0	0
			6	7	0	200	0	0	0
			7	6	0	200	0	0	0

(Continued)

TABLE A.1 (CONTINUED)

A Simulated Representative Sample for the Number of Tumors in Mice Exposed to Polycyclic Aromatic Hydrocarbons

Control	Mean	Variance	Sample	Response	Dose				
					B(a)P	B(b)F	DB(a,h)A	5MCHR	CPP
Control			8	7	0	200	0	0	0
			9	6	0	200	0	0	0
			10	8	0	200	0	0	0
			11	6	0	200	0	0	0
			12	6	0	200	0	0	0
			13	7	0	200	0	0	0
			14	7	0	200	0	0	0
			15	7	0	200	0	0	0
			16	7	0	200	0	0	0
			17	8	0	200	0	0	0
			18	7	0	200	0	0	0
			19	7	0	200	0	0	0
DB(a,h)A	1.44	2.26	1	2	0	0	1.25	0	0
			2	1	0	0	1.25	0	0
			3	1	0	0	1.25	0	0
			4	1	0	0	1.25	0	0
			5	1	0	0	1.25	0	0
			6	1	0	0	1.25	0	0
			7	1	0	0	1.25	0	0
			8	2	0	0	1.25	0	0
			9	1	0	0	1.25	0	0
			10	1	0	0	1.25	0	0

(*Continued*)

TABLE A.1 (CONTINUED)

A Simulated Representative Sample for the Number of Tumors in Mice Exposed to Polycyclic Aromatic Hydrocarbons

Control	Mean	Variance	Sample	Response	Dose				
					B(a)P	B(b)F	DB(a,h)A	5MCHR	CPP
			11	2	0	0	1.25	0	0
			12	2	0	0	1.25	0	0
			13	1	0	0	1.25	0	0
			14	2	0	0	1.25	0	0
			15	1	0	0	1.25	0	0
			16	2	0	0	1.25	0	0
			17	1	0	0	1.25	0	0
			18	2	0	0	1.25	0	0
	3.05	3.83	1	4	0	0	2.5	0	0
			2	3	0	0	2.5	0	0
			3	3	0	0	2.5	0	0
			4	3	0	0	2.5	0	0
			5	3	0	0	2.5	0	0
			6	3	0	0	2.5	0	0
			7	3	0	0	2.5	0	0
			8	4	0	0	2.5	0	0
			9	3	0	0	2.5	0	0
			10	4	0	0	2.5	0	0
			11	3	0	0	2.5	0	0
			12	3	0	0	2.5	0	0
			13	3	0	0	2.5	0	0
			14	4	0	0	2.5	0	0

(Continued)

TABLE A.1 (CONTINUED)

A Simulated Representative Sample for the Number of Tumors in Mice Exposed to Polycyclic Aromatic Hydrocarbons

Control	Mean	Variance	Sample	Response	B(a)P	B(b)F	DB(a,h)A	5MCHR	CPP
							Dose		
			15	4	0	0	2.5	0	0
			16	3	0	0	2.5	0	0
			17	3	0	0	2.5	0	0
			18	3	0	0	2.5	0	0
			19	3	0	0	2.5	0	0
	13.1	37.7	1	13	0	0	5	0	0
			2	15	0	0	5	0	0
			3	14	0	0	5	0	0
			4	12	0	0	5	0	0
			5	13	0	0	5	0	0
			6	13	0	0	5	0	0
			7	12	0	0	5	0	0
			8	15	0	0	5	0	0
			9	11	0	0	5	0	0
			10	13	0	0	5	0	0
			11	12	0	0	5	0	0
			12	13	0	0	5	0	0
			13	13	0	0	5	0	0
			14	14	0	0	5	0	0
			15	12	0	0	5	0	0
			16	12	0	0	5	0	0
			17	12	0	0	5	0	0

(Continued)

TABLE A.1 (CONTINUED)

A Simulated Representative Sample for the Number of Tumors in Mice Exposed to Polycyclic Aromatic Hydrocarbons

Control	Mean	Variance	Sample	Response	Dose				
					B(a)P	B(b)F	DB(a,h)A	5MCHR	CPP
			18	14	0	0	5	0	0
			19	15	0	0	5	0	0
			20	13	0	0	5	0	0
	28.7	123.5	1	31	0	0	10	0	0
			2	26	0	0	10	0	0
			3	31	0	0	10	0	0
			4	27	0	0	10	0	0
			5	31	0	0	10	0	0
			6	29	0	0	10	0	0
			7	32	0	0	10	0	0
			8	26	0	0	10	0	0
			9	27	0	0	10	0	0
			10	28	0	0	10	0	0
			11	30	0	0	10	0	0
			12	30	0	0	10	0	0
			13	26	0	0	10	0	0
			14	30	0	0	10	0	0
			15	31	0	0	10	0	0
			16	29	0	0	10	0	0
			17	30	0	0	10	0	0
			18	28	0	0	10	0	0
			19	31	0	0	10	0	0

(Continued)

TABLE A.1 (CONTINUED)

A Simulated Representative Sample for the Number of Tumors in Mice Exposed to Polycyclic Aromatic Hydrocarbons

Control	Mean	Variance	Sample	Response	B(a)P	B(b)F	DB(a,h)A	5MCHR	CPP
							Dose		
			20	27	0	0	10	0	0
			21	27	0	0	10	0	0
			22	30	0	0	10	0	0
			23	31	0	0	10	0	0
			24	30	0	0	10	0	0
			25	28	0	0	10	0	0
			26	31	0	0	10	0	0
			27	26	0	0	10	0	0
			28	27	0	0	10	0	0
			29	30	0	0	10	0	0
			30	30	0	0	10	0	0
5MCHR	1.75	2.62	1	2	0	0	0	10	0
			2	2	0	0	0	10	0
			3	2	0	0	0	10	0
			4	2	0	0	0	10	0
			5	2	0	0	0	10	0
			6	1	0	0	0	10	0
			7	2	0	0	0	10	0
			8	1	0	0	0	10	0
			9	2	0	0	0	10	0
			10	2	0	0	0	10	0
			11	1	0	0	0	10	0

(Continued)

TABLE A.1 (CONTINUED)

A Simulated Representative Sample for the Number of Tumors in Mice Exposed to Polycyclic Aromatic Hydrocarbons

Control	Mean	Variance	Sample	Response	B(a)P	B(b)F	DB(a,h)A	5MCHR	CPP
							Dose		
			12	1	0	0	0	10	0
			13	1	0	0	0	10	0
			14	2	0	0	0	10	0
			15	2	0	0	0	10	0
			16	2	0	0	0	10	0
			17	2	0	0	0	10	0
			18	1	0	0	0	10	0
			19	1	0	0	0	10	0
			20	1	0	0	0	10	0
	13.5	54.5	1	13	0	0	0	30	0
			2	16	0	0	0	30	0
			3	12	0	0	0	30	0
			4	12	0	0	0	30	0
			5	14	0	0	0	30	0
			6	14	0	0	0	30	0
			7	12	0	0	0	30	0
			8	15	0	0	0	30	0
			9	13	0	0	0	30	0
			10	12	0	0	0	30	0
	39	187.4	1	36	0	0	0	50	0
			2	42	0	0	0	50	0
			3	41	0	0	0	50	0

(Continued)

TABLE A.1 (CONTINUED)

A Simulated Representative Sample for the Number of Tumors in Mice Exposed to Polycyclic Aromatic Hydrocarbons

Control	Mean	Variance	Sample	Response	Dose				
					B(a)P	B(b)F	DB(a,h)A	5MCHR	CPP
			4	42	0	0	0	50	0
			5	36	0	0	0	50	0
			6	38	0	0	0	50	0
			7	41	0	0	0	50	0
			8	40	0	0	0	50	0
			9	36	0	0	0	50	0
			10	35	0	0	0	50	0
			11	41	0	0	0	50	0
			12	39	0	0	0	50	0
			13	42	0	0	0	50	0
			14	35	0	0	0	50	0
			15	41	0	0	0	50	0
			16	40	0	0	0	50	0
			17	38	0	0	0	50	0
			18	42	0	0	0	50	0
	93.1	398.6	1	98	0	0	0	100	0
			2	97	0	0	0	100	0
			3	96	0	0	0	100	0
			4	98	0	0	0	100	0
			5	93	0	0	0	100	0
			6	96	0	0	0	100	0
			7	89	0	0	0	100	0

(Continued)

TABLE A.1 (CONTINUED)

A Simulated Representative Sample for the Number of Tumors in Mice Exposed to Polycyclic Aromatic Hydrocarbons

Control	Mean	Variance	Sample	Response	Dose				
					B(a)P	B(b)F	DB(a,h)A	5MCHR	CPP
			8	89	0	0	0	100	0
			9	95	0	0	0	100	0
			10	90	0	0	0	100	0
			11	88	0	0	0	100	0
			12	98	0	0	0	100	0
			13	98	0	0	0	100	0
			14	93	0	0	0	100	0
			15	96	0	0	0	100	0
CPP	0.55	0.682	1	1	0	0	0	0	10
			2	1	0	0	0	0	10
			3	1	0	0	0	0	10
			4	0	0	0	0	0	10
			5	1	0	0	0	0	10
			6	1	0	0	0	0	10
			7	1	0	0	0	0	10
			8	1	0	0	0	0	10
			9	1	0	0	0	0	10
			10	1	0	0	0	0	10
			11	1	0	0	0	0	10
			12	1	0	0	0	0	10
			13	1	0	0	0	0	10
			14	0	0	0	0	0	10

(Continued)

TABLE A.1 (CONTINUED)

A Simulated Representative Sample for the Number of Tumors in Mice Exposed to Polycyclic Aromatic Hydrocarbons

Control	Mean	Variance	Sample	Response	Dose				
					B(a)P	B(b)F	DB(a,h)A	5MCHR	CPP
			15	1	0	0	0	0	10
			16	0	0	0	0	0	10
			17	0	0	0	0	0	10
			18	0	0	0	0	0	10
			19	0	0	0	0	0	10
			20	1	0	0	0	0	10
	4.75	4.72	1	5	0	0	0	0	50
			2	4	0	0	0	0	50
			3	5	0	0	0	0	50
			4	5	0	0	0	0	50
			5	4	0	0	0	0	50
			6	4	0	0	0	0	50
			7	4	0	0	0	0	50
			8	4	0	0	0	0	50
			9	4	0	0	0	0	50
			10	5	0	0	0	0	50
			11	5	0	0	0	0	50
			12	5	0	0	0	0	50
			13	5	0	0	0	0	50
			14	5	0	0	0	0	50
			15	5	0	0	0	0	50
			16	5	0	0	0	0	50

(*Continued*)

TABLE A.1 (CONTINUED)

A Simulated Representative Sample for the Number of Tumors in Mice Exposed to Polycyclic Aromatic Hydrocarbons

Control	Mean	Variance	Sample	Response	Dose				
					B(a)P	B(b)F	DB(a,h)A	5MCHR	CPP
			17	5	0	0	0	0	50
			18	5	0	0	0	0	50
			19	4	0	0	0	0	50
			20	5	0	0	0	0	50
	29.25	215.7	1	27	0	0	0	0	100
			2	31	0	0	0	0	100
			3	30	0	0	0	0	100
			4	30	0	0	0	0	100
			5	31	0	0	0	0	100
			6	30	0	0	0	0	100
			7	26	0	0	0	0	100
			8	26	0	0	0	0	100
			9	29	0	0	0	0	100
			10	33	0	0	0	0	100
			11	33	0	0	0	0	100
			12	31	0	0	0	0	100
			13	30	0	0	0	0	100
			14	33	0	0	0	0	100
			15	29	0	0	0	0	100
			16	31	0	0	0	0	100
			17	32	0	0	0	0	100
			18	32	0	0	0	0	100

(Continued)

TABLE A.1 (CONTINUED)

A Simulated Representative Sample for the Number of Tumors in Mice Exposed to Polycyclic Aromatic Hydrocarbons

Control	Mean	Variance	Sample	Response	B(a)P	B(b)F	DB(a,h)A	5MCHR	CPP
			19	29	0	0	0	0	100
			20	30	0	0	0	0	100
			21	29	0	0	0	0	100
			22	29	0	0	0	0	100
			23	26	0	0	0	0	100
			24	28	0	0	0	0	100
			25	26	0	0	0	0	100
			26	28	0	0	0	0	100
			27	32	0	0	0	0	100
			28	32	0	0	0	0	100
	97.7	868.1	1	91	0	0	0	0	200
			2	93	0	0	0	0	200
			3	99	0	0	0	0	200
			4	89	0	0	0	0	200
			5	98	0	0	0	0	200
			6	103	0	0	0	0	200
			7	102	0	0	0	0	200
			8	92	0	0	0	0	200
			9	92	0	0	0	0	200
			10	106	0	0	0	0	200
			11	92	0	0	0	0	200
			12	91	0	0	0	0	200

(Continued)

TABLE A.1 (CONTINUED)

A Simulated Representative Sample for the Number of Tumors in Mice Exposed to Polycyclic Aromatic Hydrocarbons

Control	Mean	Variance	Sample	Response	Dose					
					B(a)P	B(b)F	DB(a,h)A	5MCHR	CPP
			13	98	0	0	0	0	200
			14	105	0	0	0	0	200
			15	102	0	0	0	0	200
			16	90	0	0	0	0	200
			17	98	0	0	0	0	200
			18	98	0	0	0	0	200
			19	102	0	0	0	0	200

References

1. Gennings, C. (1995). An efficient experimental design for detecting departure from additivity in mixtures of many chemicals. *Toxicology*, 10, 189–97.
2. Nesnow, S., Mass, M. J., Ross, J. A., Galati, A. J., Lambert, G. R., Gennings, C., Carter, W. H. Jr., and Stoner, G. D. (1998). Lung tumorigenic interactions in strain A/J mice of five environmental polycyclic aromatic hydrocarbons. *Environmental Health Perspectives*, 106, 1337–46.

Reference

Appendix 2: Data Generation for a Mixture of Four Metals (Example 3.5)

In Chapter 3, we presented a methodology by Casey et al. (2004) to assess the additivity of the effect of a mixture of chemicals. To illustrate the application of the methodology, in Example 3.5, we examined the additivity of the effect of a mixture of four metals: arsenic, chromium, cadmium, and lead. Using the summary data in Gennings et al. (2002), for the purpose of illustration, it was necessary to generate a representative sample outcome of interest, percent viability of treated normal human epidermal keratino-cytes (NHEK) cells using 3-45-Dimethylthiazol-2-Yl-25-Diphenyltetrazolium (MTT) assay. Here, we describe the methodology for generating that representative data set.

Normal random variates with given means and standard deviation were generated in Excel and rounded to the nearest integer by using the function ROUND(NORMINV(RAND(),C2,SQRT(D2)),0). The process was repeated for the given sample size. Tables A.2 and A.3 give the simulated representative data for individual chemicals and for the mixture respectively.

TABLE A.2

Generated Responses for each Chemical

Response	Arsenic	Chromium	Cadmium	Lead
		Dosage		
111	0	0	0	0
99	0	0	0	0
102	0	0	0	0
89	0	0	0	0
100	0	0	0	0
102	0	0	0	0
99	0	0	0	0
100	0	0	0	0
86	0	0	0	0
98	0.3	0	0	0
98	0.3	0	0	0
117	0.3	0	0	0
128	0.3	0	0	0
106	0.3	0	0	0
105	0.3	0	0	0
136	0.3	0	0	0
120	0.3	0	0	0
129	0.3	0	0	0
100	1	0	0	0
97	1	0	0	0
103	1	0	0	0
85	1	0	0	0
109	1	0	0	0
98	1	0	0	0
94	1	0	0	0
112	1	0	0	0
75	1	0	0	0
54	3	0	0	0
66	3	0	0	0
70	3	0	0	0
102	3	0	0	0
96	3	0	0	0
81	3	0	0	0
92	3	0	0	0
96	3	0	0	0
85	3	0	0	0
70	10	0	0	0
32	10	0	0	0
64	10	0	0	0

(Continued)

TABLE A.2 (CONTINUED)

Generated Responses for each Chemical

	Dosage			
Response	Arsenic	Chromium	Cadmium	Lead
21	10	0	0	0
41	10	0	0	0
39	10	0	0	0
48	10	0	0	0
49	10	0	0	0
59	10	0	0	0
−4	30	0	0	0
22	30	0	0	0
16	30	0	0	0
10	30	0	0	0
17	30	0	0	0
14	30	0	0	0
−8	30	0	0	0
18	30	0	0	0
14	30	0	0	0
98	0	0	0	0
103	0	0	0	0
109	0	0	0	0
101	0	0	0	0
85	0	0	0	0
105	0	0	0	0
110	0	0.3	0	0
89	0	0.3	0	0
114	0	0.3	0	0
89	0	0.3	0	0
100	0	0.3	0	0
104	0	0.3	0	0
110	0	0.3	0	0
105	0	0.3	0	0
106	0	0.3	0	0
95	0	1	0	0
118	0	1	0	0
114	0	1	0	0
114	0	1	0	0
109	0	1	0	0
101	0	1	0	0
97	0	1	0	0
100	0	1	0	0
104	0	1	0	0

(Continued)

TABLE A.2 (CONTINUED)

Generated Responses for each Chemical

	Dosage			
Response	Arsenic	Chromium	Cadmium	Lead
62	0	3	0	0
64	0	3	0	0
70	0	3	0	0
83	0	3	0	0
69	0	3	0	0
104	0	3	0	0
85	0	3	0	0
77	0	3	0	0
106	0	3	0	0
49	0	10	0	0
20	0	10	0	0
−2	0	10	0	0
43	0	10	0	0
27	0	10	0	0
0	0	10	0	0
0	0	10	0	0
45	0	10	0	0
13	0	10	0	0
−10	0	30	0	0
3	0	30	0	0
−11	0	30	0	0
21	0	30	0	0
14	0	30	0	0
−11	0	30	0	0
3	0	30	0	0
9	0	30	0	0
6	0	30	0	0
99	0	0	0	0
105	0	0	0	0
97	0	0	0	0
96	0	0	0	0
92	0	0	0	0
97	0	0	0	0
118	0	0	3	0
75	0	0	3	0
106	0	0	3	0
85	0	0	3	0
94	0	0	3	0
97	0	0	3	0

(Continued)

TABLE A.2 (CONTINUED)

Generated Responses for each Chemical

	Dosage			
Response	Arsenic	Chromium	Cadmium	Lead
99	0	0	3	0
107	0	0	3	0
100	0	0	3	0
29	0	0	10	0
74	0	0	10	0
63	0	0	10	0
43	0	0	10	0
82	0	0	10	0
81	0	0	10	0
20	0	0	10	0
113	0	0	10	0
103	0	0	10	0
22	0	0	30	0
−2	0	0	30	0
43	0	0	30	0
53	0	0	30	0
27	0	0	30	0
46	0	0	30	0
50	0	0	30	0
−1	0	0	30	0
−2	0	0	30	0
−2	0	0	100	0
4	0	0	100	0
1	0	0	100	0
0	0	0	100	0
6	0	0	100	0
3	0	0	100	0
2	0	0	100	0
0	0	0	100	0
6	0	0	100	0
3	0	0	300	0
11	0	0	300	0
10	0	0	300	0
−2	0	0	300	0
6	0	0	300	0
5	0	0	300	0
9	0	0	300	0
5	0	0	300	0
−1	0	0	300	0

(Continued)

TABLE A.2 (CONTINUED)

Generated Responses for each Chemical

	Dosage			
Response	Arsenic	Chromium	Cadmium	Lead
98	0	0	0	0
100	0	0	0	0
98	0	0	0	0
100	0	0	0	0
98	0	0	0	0
101	0	0	0	0
92	0	0	0	3
96	0	0	0	3
86	0	0	0	3
80	0	0	0	3
87	0	0	0	3
108	0	0	0	3
95	0	0	0	3
94	0	0	0	3
100	0	0	0	3
97	0	0	0	10
96	0	0	0	10
83	0	0	0	10
105	0	0	0	10
78	0	0	0	10
96	0	0	0	10
100	0	0	0	10
71	0	0	0	10
89	0	0	0	10
112	0	0	0	30
109	0	0	0	30
96	0	0	0	30
110	0	0	0	30
85	0	0	0	30
86	0	0	0	30
107	0	0	0	30
95	0	0	0	30
106	0	0	0	30
178	0	0	0	100
139	0	0	0	100
−26	0	0	0	100
69	0	0	0	100
76	0	0	0	100

(Continued)

TABLE A.2 (CONTINUED)

Generated Responses for each Chemical

	Dosage			
Response	Arsenic	Chromium	Cadmium	Lead
97	0	0	0	100
52	0	0	0	100
53	0	0	0	100
104	0	0	0	100
20	0	0	0	300
21	0	0	0	300
35	0	0	0	300
5	0	0	0	300
−1	0	0	0	300
21	0	0	0	300
29	0	0	0	300
24	0	0	0	300
25	0	0	0	300

TABLE A.3

Generated Responses for the Mixture of Four Chemicals

Dose	Sample #	Response
0.16	1	122
0.16	2	126
0.16	3	120
0.16	4	113
0.16	5	138
0.16	6	134
0.16	7	98
0.16	8	135
0.16	9	74
0.48	1	97
0.48	2	94
0.48	3	94
0.48	4	97
0.48	5	94
0.48	6	98
0.48	7	93
0.48	8	94
0.48	9	95
1.46	1	89
1.46	2	111
1.46	3	84

(Continued)

TABLE A.3 (CONTINUED)

Generated Responses for the Mixture of Four Chemicals

Dose	Sample #	Response
1.46	4	60
1.46	5	86
1.46	6	91
1.46	7	70
1.46	8	83
1.46	9	93
4.39	1	71
4.39	2	77
4.39	3	89
4.39	4	78
4.39	5	88
4.39	6	86
4.39	7	73
4.39	8	75
4.39	9	89
13.2	1	69
13.2	2	65
13.2	3	65
13.2	4	65
13.2	5	66
13.2	6	71
13.2	7	61
13.2	8	62
13.2	9	63
39.6	1	50
39.6	2	41
39.6	3	54
39.6	4	41
39.6	5	49
39.6	6	62
39.6	7	45
39.6	8	30
39.6	9	31
118.7	1	37
118.7	2	28
118.7	3	23
118.7	4	39
118.7	5	37
118.7	6	34
118.7	7	50
118.7	8	56
118.7	9	44

References

1. Casey, M., Gennings, C., Carter, W. H. Jr., Moser, V., and Simmons, J. E. (2004). Detecting interaction and assessing the impact component subset in a chemical mixture using fixed-ratio ray designs. *Journal of Agricultural, Biological, and Environmental Statistics*, 9, 339–61.
2. Gennings, C., Carter, W. H. Jr., Campain, J. A., Bae, D., and Yang, R. S. H. (2002). Statistical analysis of cytotoxicity in human epidermal keratinocytes following exposure to a mixture of four metals. *Journal of Agricultural, Biological, and Environmental Statistics*, 7, 58–73.

Appendix 3: Model Final Exam Questions

1. An animal bioassay experiment consists of g dose groups. At dose level d_i, $i=1, ..., g$, n_i animals are exposed and the number of responses Y_i is observed.

   a. Assuming a linear logistic model for the probability of response π_i, write down the likelihood for estimating the unknown parameters and derive the likelihood equations.

   b. Show that the likelihood equations can be expressed as

   $$\sum_{i=1}^{g}\left(Y_i - \hat{Y}_i\right) = 0$$

   $$\sum_{i=1}^{g} d_i\left(Y_i - \hat{Y}_i\right) = 0$$

   where $\hat{Y}_i = n_i \hat{\pi}_i$.

   c. Suppose now that $g=2$ as in case-control studies. Show that the exact solution for estimating the regression parameters β_0 and β_1 is given by

   $$\hat{\beta}_0 = \frac{d_1 A_2 - d_2 A_1}{d_1 - d_2}$$

   $$\hat{\beta}_1 = \frac{A_1 - A_2}{d_1 - d_2}$$

   where

   $$A_i = \ln\left(\frac{Y_i}{Y_i - n_i}\right); \quad i = 1, 2$$

   d. Show that the effect dose that achieves 50% response is given by

   $$ED_{50} = -\frac{\beta_0}{\beta_1}.$$

   e. In an experiment, 100 mice were randomly assigned to two equal-sized groups. One group served as control and received the placebo, while the animals in the other group were exposed to 50 mg/kg body weight benzene per day for a month. At the conclusion of the experiment, the number of rats with tumor in the control group was 1, while that in the treatment group was 20. Estimate the ED_{50} for this experiment and comment.

2. In a bioassay experiment to assess the neurotoxic effect of a chemical, suppose responses $Y_1, Y_2, ..., Y_g$ are resulted when subjects are exposed to doses $d_1, d_2, ..., d_g$ of a chemical. Assume a simple linear regression dose-response model

$$Y_i \sim N\left(\beta_0 + \beta_1 d_i, \sigma^2\right); \quad i = 1, ..., g$$

Suppose that an abnormal adverse effect is defined as a response that falls more than k standard deviations below the control mean for a fixed $k > 0$. Let $\pi(d)$, the measure of risk at dose d, be defined as the additional risk over the background.

a. Show that

$$\pi(d) = \Phi(-\omega d - k) - \Phi(-k) \tag{A.1}$$

where $\Phi(.)$ is the standard normal cdf and $\omega = \dfrac{\beta_1}{\sigma}$.

b. Using the standard linear regression theory, give the maximum likelihood estimates $\hat{\beta}_1$ and $\hat{\omega}$ for β_1 and ω respectively.

c. Show that while $\hat{\beta}_1$ is unbiased, $\hat{\omega}$ is biased, and

$$E(\hat{\omega}) = \frac{\Gamma\left(\dfrac{g-3}{2}\right) \cdot \sqrt{\dfrac{g-2}{2}}}{\Gamma\left(\dfrac{g-2}{2}\right)} \omega$$

for $g > 3$, where $\Gamma(.)$ is the gamma function.

d. Using the result of part c., define an unbiased estimate $\tilde{\omega}$ for ω.

e. Use $\tilde{\omega}$ in (A.1) and derive the benchmark dose (BMD) for a given benchmark risk (BMR) and briefly explain how we can calculate the BMD lower confidence limit (BMDL).

f. Estimate the no-observed-adverse-effect-level (NOAEL) and the 5% BMD for the following data set:

Adult female rhesus monkeys (three per dose group) were exposed to 0, 5.0, or 10.0 mg/kg of methylenedioxymethamphetamine (MDMA) for four consecutive days and the concentration of neurotransmitter serotonin (5-HT) in a specific region of the brain was measured after five weeks. The results are presented in Table A.4:

3. The attached data set consists of the litter weights of a developmental toxicity study by Dempster et al. (1984).

   a. Apply an appropriate statistical test to see if the litter weights for each sex can be assumed to have a normal distribution.

   b. If necessary, apply an appropriate transformation, e.g. logarithmic or arcsine, and test for normality.

   c. It is thought, in general, that male pups on average weigh more than the female pups. Verify this hypothesis assuming normal distributions for the fetal weights.

   d. From figure 1 in Dempster et al. (1984), it appears that the average litter weight decreases as the litter size increases. Use linear regression to determine a linear predictor of the mean litter weight as a function of the litter size. Test for the significance of the model. Is the same true for each dose group?

   e. Use a simple one-way analysis of variance to verify if there is a significant difference in the mean fetal weights of the three dose groups. If significant, apply a *post-hoc* test such as William's test or Dunnett's test to determine the NOAEL. Comment on the appropriateness of the model.

   f. Apply a two-level nested model to the litter weights and test for model significance. Give estimates of the variance components and the intralitter correlation.

   g. Now, apply a mixed-effects model with a random effect due to litter variation to the mean litter weights. Test for the significance of the model and give estimates for the variance components. Also give an estimate of the intralitter correlation.

   h. Repeat the analysis using sex and litter size as covariates and draw conclusions (Table A.5).

4. In a developmental toxicity experiment with g dose levels, m_i, $i=1$, ..., g, pregnant female animals are exposed to a dose d_i of a chemical. Let π_1 be the probability of a malformation in a viable fetus and

TABLE A.4

Neurotoxicity of MDMA in Monkey

Dose	0	5	10
5-HT concentration	4.5	0.8	1.0
	3.8	1.4	0.5
	4.3	1.1	0.6

Source: Data from Gaylor and Slikker (1990).

TABLE A.5

Fetal Weight by Sex

	Dam	Males	Females
Controls	1	6.60	6.95
		7.40	6.29
		7.15	6.77
		7.24	6.57
		7.10	
		6.04	
		6.98	
		7.05	
	2	6.37	5.92
		6.37	6.04
		6.90	5.82
		6.34	6.04
		6.50	5.96
		6.10	
		6.44	
		6.94	
		6.41	
	3	7.50	7.57
		7.08	7.27
	4	6.25	6.29
		6.93	5.98
		6.80	6.32
		6.69	6.28
		6.28	5.65
		6.27	5.57
		6.27	
		6.47	
	5	7.96	7.16
		6.84	7.09
		7.00	7.14
		8.10	5.02
		6.52	6.04
		7.23	
		6.10	
		7.31	
	6	8.26	7.26
		7.73	6.58
		8.33	3.68
		6.14	
		7.75	
		6.96	

(Continued)

TABLE A.5 (CONTINUED)

Fetal Weight by Sex

	Dam	Males	Females
	7	6.29	6.16
		6.32	5.96
		6.28	6.26
		6.24	5.83
		6.78	6.11
		6.63	6.45
		6.27	6.25
		6.29	6.31
		6.06	5.74
	8	6.04	6.23
		5.84	5.95
		6.77	6.16
		5.59	6.19
		5.52	5.32
		6.42	5.00
		5.97	6.30
		6.34	5.00
			5.56
	9	5.37	5.37
		5.58	5.33
		5.51	5.44
		5.19	5.14
		5.34	
		5.77	
		5.17	
		4.57	
		5.39	
		5.62	
		5.40	
		5.77	
		5.24	
	10	7.30	6.44
		6.60	6.67
		6.58	6.43
		6.68	6.53
		6.46	5.92
		6.38	6.52
			6.44
Low dose	11	5.65	5.54
		5.78	5.72
		6.23	5.50

(*Continued*)

TABLE A.5 (CONTINUED)

Fetal Weight by Sex

Dam	Males	Females
	5.70	5.64
	5.73	5.42
	6.10	5.42
	5.55	
	5.71	
	5.81	
	6.10	
12		6.89
		7.73
13	5.83	6.09
	5.97	5.39
	6.39	5.89
	5.69	5.14
	5.69	
	5.97	
	6.04	
	5.46	
14	5.92	5.66
	5.75	5.76
	6.22	5.73
	5.96	5.33
	5.59	5.58
	5.79	5.88
	6.23	
	5.88	
	6.02	
15	6.00	5.96
	6.11	6.32
	6.40	5.83
	6.06	5.97
	6.39	5.87
	6.09	5.67
	6.32	
16	6.43	6.09
	6.13	5.63
	5.87	5.84
		6.20
		6.42
		5.90
		5.62
		6.23

(Continued)

TABLE A.5 (CONTINUED)

Fetal Weight by Sex

	Dam	Males	Females
			5.85
			5.89
	17	5.81	5.63
		5.44	5.12
		5.65	5.65
		5.25	5.29
		5.45	5.13
		5.32	5.60
		5.89	5.08
	18	6.77	6.49
		7.13	6.09
		6.85	6.09
			5.99
			6.11
			6.15
			4.75
			5.69
			6.19
			5.72
			6.14
	19	6.72	6.11
		6.34	5.71
		6.48	6.41
		5.74	6.21
			6.11
			5.81
	20	5.90	6.12
		6.22	5.40
		6.67	5.50
		6.23	5.46
		6.24	5.97
		6.26	6.11
		6.36	
		6.05	
		5.89	
		6.29	
High dose	21	5.09	5.23
		5.57	5.13
		5.69	4.48
		5.50	
		5.45	

(Continued)

TABLE A.5 (CONTINUED)

Fetal Weight by Sex

	Dam	Males	Females
		5.24	
		5.36	
		5.26	
		5.36	
		5.01	
		5.03	
	22	5.30	5.19
		5.40	5.42
		5.55	5.40
		6.02	5.12
		5.27	5.40
	23	7.70	7.68
			6.33
	24	6.28	5.68
		5.74	5.76
		6.29	6.03
			5.30
			5.55
			6.53
			5.76
			5.77
			5.49
	25	6.50	6.10
		7.10	6.63
		7.00	6.33
		7.00	
		5.85	
	26	7.00	6.22
		6.15	6.20
			5.76
			6.21
			6.42
			6.42
			6.30
	27	5.64	5.93
		6.06	5.74
		6.56	5.74
		6.29	
		5.69	
		6.36	

Source: Data from Dempster et al. (1984).

let π_2 be the probability of death/resorption. Denote the intralitter correlation by φ. Then the probability of a malformation given that the fetus is viable is $\mu = \pi_1(1 - \pi_2)$. Let $\mu_2 = \pi_2$ and denote the vector of responses by $Z = (Y, R)^T$ where Y is the number of malformations and R is the number of deaths/resorptions.

a. Using standard multinomial results, if n is the number of implants, find $\mu = (Z|n)$ and $V = Cov(Z|n)$.

b. Let n_{ij} be the number of implants for the jth litter of the ith dose group and denote by y_{ij}, r_{ij} and s_{ij} the number of malformed, dead/resorbed, and live fetuses respectively.

(Note $s_{ij} - y_{ij}$ is the number of healthy fetuses.) Let $Z_{ij} = (Y_{ij}, R_{ij})^T$; $j = 1, \ldots, m_i, i = 1, \ldots, g$ be the vector of outcomes for the jth litter of the ith dose group. Denote the mean function for dose d_i by $(d_i, \theta) = (\mu_{i1}, \mu_{i2})$ where $\theta = \left(\theta_1^T, \theta_2^T\right)^T$ is the vector of parameters. Let $W = [1 + (n - 1) \varphi] V$. (Note that W corresponds to the covariance matrix of a trinomial distribution when $\varphi = 0$ and a Dirichlet-trinomial distribution when $0 \leq \varphi \leq 1$.) Using the covariance matrix you derived in part a. as the working covariance, derive the generalized estimating equations (GEEs) for estimating θ_1 and θ_2 and show that they lead to two separate sets of equations, one in terms of y_{ij} and θ_1, and the other in terms of s_{ij} and θ_2 (Zhu et al., 1994).

Hint: The GEEs are given by

$$\sum_{i=1}^{g}\sum_{j=1}^{m_i} n_{ij} D_i^T W_{ij}^T \left(Z_{ij} - n_{ij}\mu_i\right) = 0$$

where $D_i = \dfrac{\partial \mu_i}{\partial \theta}$ and W_{ij} = working covariance matrices.

5. A compound consists of a mixture of two chemicals. For the dose d_1 of chemical 1 and dose d_2 of chemical 2, Berenbaum (1985) suggests characterizing the interaction of the two chemicals using an "interaction index" defined by

$$I = \frac{d_1}{ED_{100\alpha}^{(1)}} + \frac{d_2}{ED_{100\alpha}^{(2)}}$$

where $ED_{100\alpha}^{(k)}$ is the effective dose corresponding to a response level of $100\,\alpha\%$ for chemical k; $k = 1, 2$.

a. Using a linear logistic dose-response model for each chemical, derive an expression for the interaction index.

b. Now, suppose that the joint action of the two chemicals is expressed by a quadratic model as

$$g(d_1, d_2) = \beta_0 + \beta_1 d_1 + \beta_2 d_2 + \beta_3 d^2 + \beta_4 d^2 + \beta_5 d_1 d_2$$

where g is an appropriate link function. Let the measure of risk for the joint action be defined as the extra risk

$$\pi(d_1, d_2) = \frac{g(d_1, d_2) - g(0,0)}{1 - g(0,0)}.$$

Given a response level π for the joint action, derive an expression for the benchmark profile of chemical 1 for a fixed dose d_2 of chemical 2. Illustrate for the complementary log link function.

References

1. Berenbaum, M. C. (1985). The expected effect of combination of agents: The general solution. *Journal of Theoretical Biology*, 114, 413–31.
2. Dempster, A. P., Selwyn, M. R., Patel, C. M., and Roth, A. J. (1984). Statistical and computational aspects of mixed model analysis. *Applied Statistics*, 33, 203–14.
3. Gaylor, D. W., and Slikker, J. W. (1990). Risk assessment for neurotoxic effects. *Neurotoxicology*, 11(2), 211–18.
4. Zhu, Y., Krewski, D., and Ross, W. H. (1994). Dose-response models for correlated multinomial data from developmental toxicity studies. *Applied Statistics*, 43, 583–98.

Index

Printed in the United States
by Baker & Taylor Publisher Services